Ben Stacy Jerrik (Ed.)

Navarretia Pubescens

Ben Stacy Jerrik (Ed.)

# Navarretia Pubescens

Polemoniaceae, Navarretia, Pubescens,
Pincushionplant, Habitat, Flowers

Part Press

# Wayne Bolt

# American_football

The U.S. Naval Academy Midshipmen (left) face off against the Colorado State Rams.

| Nickname(s) | Gridiron |
| --- | --- |
| First played | November 6, 1869, Rutgers vs. Princeton |
| Characteristics | |
| Contact | Full contact |
| Team members | 11 at a time |
| Categorization | Outdoor |
| Equipment | Football |
| Olympic | No |

**American football** is a sport played between two teams of eleven with the objective of scoring points by advancing the ball into the opposing team's end zone. Known in the United States simply as **football**, it may also be referred to informally as **gridiron football**.[1] [2] The ball can be advanced by running with it or throwing it to a teammate. Points can be scored by carrying the ball over the opponent's goal line, catching a pass thrown over that goal line, kicking the ball through the opponent's goal posts or tackling an opposing ball carrier in his own end zone.

In the United States, the major forms are high school football, college football and professional football. Each of these are played under slightly different rules.[3] High school football is governed by the National Federation of State High School Associations and college football by the National Collegiate Athletic Association and National Association of Intercollegiate Athletics. The major league for professional football is the National Football League.

American football is closely related to Canadian football but with some differences in rules and the field.[4] Both sports can be traced to early versions of association football and rugby football.

# History

The history of American football can be traced to early versions of rugby football and association football. Both games have their origins in varieties of football played in the United Kingdom in the mid-19th century, in which a ball is kicked at a goal and/or run over a line. Many games known as "football" were being played at colleges and universities in the United States in the first half of the 19th century.[5][6] Modern American football grew off a historical game played between Harvard and McGill University in 1874.

Rutgers University and its neighbor, Princeton University, played the first game of intercollegiate football on 6 November 1869 on a plot of ground where the present-day Rutgers gymnasium now stands in New Brunswick, N.J. Rutgers won that first game, 6-4.[7]

An American football team at the turn of the 20th century. Note the continued use of a rugby-type ball (front row, right).

Walter Camp

American football resulted from several major divergences from rugby football, most notably the rule changes instituted by Walter Camp, considered the "Father of American Football." Among these important changes were the introduction of the line of scrimmage and of down-and-distance rules. In the late 19th and early 20th centuries, game play developments by college coaches such as Eddie Cochems, Amos Alonzo Stagg, Knute Rockne, and Glenn "Pop" Warner helped take advantage of the newly introduced forward pass.

The popularity of **collegiate football** grew as it became the dominant version of the sport for the first half of the twentieth century. Bowl games, a college football tradition, attracted a national audience for collegiate teams. Bolstered by fierce rivalries, college football still holds widespread appeal in the US.[5] [8] [9]

The origin of **professional football** can be traced back to 1892, with William "Pudge" Heffelfinger's $500 contract to play in a game for the Allegheny Athletic Association against the Pittsburgh Athletic Club. The first Professional "league" was the Ohio League, formed in 1903, and the first Professional Football championship game was between the Buffalo Prospects and the Canton Bulldogs in 1919. In 1920, the American Professional Football Association was formed. The first game was played in Dayton, Ohio on October 3, 1920 with the host Triangles defeating the Columbus Panhandles 14–0. The league changed its name to the National Football League (NFL) two years later, and eventually became the major league of American football. Initially a sport of Midwestern industrial towns in the United States, professional football eventually became a national phenomenon. Football's increasing popularity is usually traced to the 1958 NFL Championship Game, a contest that has been dubbed the "Greatest Game Ever Played". A rival league to the NFL, the American Football League (AFL), began play in 1960; the pressure it put on the senior league led to a merger between the two leagues and the creation of the Super Bowl, which has become the most watched television event in the United States on an annual basis.[10]

**High school football** dates to the 1880s and enjoys regional popularity.

# Rules

## Field and players

American football is played on a field 360 by 160 feet (120.0 by 53.3 yards; 109.7 by 48.8 meters).[11] The longer boundary lines are *sidelines*, while the shorter boundary lines are *end lines*. Sidelines and end lines are out of bounds. Near each end of the field is a *goal line*; they are 100 yards (91.4 m) apart. A scoring area called an *end zone* extends 10 yards (9.1 m) beyond each goal line to each end line. The end zone includes the goal line but not the end line.[11] While the playing field is effectively flat, it is common for a field to be built with a slight crown—with the middle of the field higher than the sides—to allow water to drain from the field.

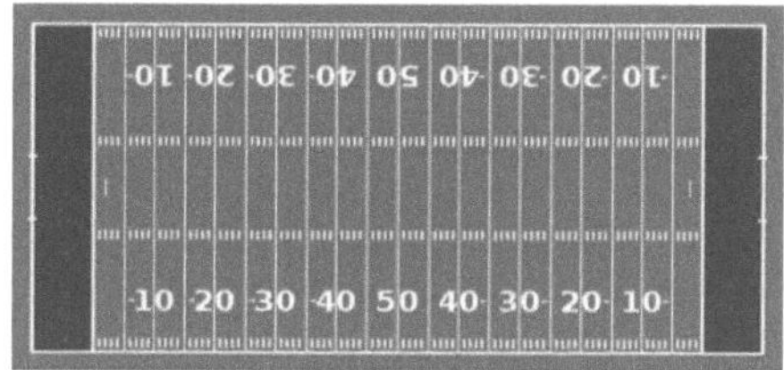

The numbers on the field indicate the number of yards to the nearest end zone.

*Yard lines* cross the field every 5 yards (4.6 m), and are numbered every 10 yards from each goal line to the 50-yard line, or midfield (similar to a typical rugby league field). Two rows of short lines, known as inbounds lines or *hash marks*, run at 1-yard (91.4 cm) intervals perpendicular to the sidelines near the middle of the field. All plays start with the ball on or between the hash marks. Because of the arrangement of the lines, the field is occasionally referred to as a *gridiron* in a reference to the cooking grill with a similar pattern of lines.

At the back of each end zone are two *goalposts* (also called *uprights*) connected by a crossbar 10 feet (3.05 m) from the ground. For high skill levels, the posts are 18 feet 6 inches (5.64 m) apart. For lower skill levels, these are widened to 23 feet 4 inches (7.11 m).

Each team has 11 players on the field at a time. Usually there are many more players off the field (an NFL team has a limit of 53 players on their roster, 46 of which can be dressed for a game). However, teams may substitute for any or all of their players during the breaks between plays. As a result, players have very specialized roles and are divided into three separate units: the offense, the defense and the special teams. It is rare for all team members to participate in a given game, as some roles have little utility beyond that of an injury substitute.

## Start of halves

The game begins with a coin toss to determine which team will kick off to begin the game and which goal each team will defend.[12] The options are presented again to start the second half; the choices for the first half do not automatically determine the start of the second half. The referee conducts the coin toss with the captains (or sometimes coaches) of the opposing teams. The team that wins the coin toss has three options:[12]

1. They may choose whether to kick or receive the opening kickoff.
2. They may choose which goal to defend.
3. They may choose to *defer* the first choice to the other team and have first choice to start the second half.[13]

Whatever the first team chooses, the second team has the option on the other choice (for example, if the first team elects to receive at the start of the game, the second team can decide which goal to defend).

At the start of the second half, the options to kick, receive, or choose a goal to defend are presented to the captains again. The team which did not choose first to start the first half (or which deferred its privilege to choose first) now gets first choice of options.[12] [14]

## Game duration

A standard football game consists of four 15-minute quarters (12-minute quarters in high-school football and often shorter at lower levels),[15] with a half-time intermission after the second quarter.[16] Depending upon the level of competition, the duration of the half-time ranges from 10 to 20 minutes. At all levels, a down (play) that begins before time expires is allowed to continue until its completion, even after the clock reaches zero. The clock is also stopped after certain plays, therefore, a game can last considerably longer (often more than three hours in real time), and if a game is broadcast on television, TV timeouts are taken at certain intervals of the game to broadcast commercials outside of game action. If an NFL game is tied after four quarters, the teams play an additional period lasting up to 15 minutes. In a regular season NFL overtime game, the first team that scores wins, even if the other team does not get a possession; this is referred to as sudden death. However, in a post-season NFL game during the playoffs, if the first team with possession scores only a field goal, the other team is allowed the opportunity to match or better this score. This rule only affects playoff games in overtime in which the first team with possession scores a field goal: if the first team with possession scores a touchdown, the sudden death rules take effect. In a regular-season NFL game, if neither team scores in overtime, the game is a tie. In an NFL playoff game, additional overtime periods are played, as needed, to determine a winner. College overtime rules are more complicated.

## Advancing the ball

The team that takes possession of the ball (the **offense**) has four attempts, called **downs**, in which to advance the ball at least 10 yards toward their opponent's (the **defense**'s) end zone. When the offense succeeds in gaining at least 10 yards, it gets a **first down**, meaning the team starts a new set of four downs to gain yet another 10 yards or to score. If the offense fails to gain a first down (10 yards) after four downs, the other team gets possession of the ball at the point where the fourth down ended, beginning with their first down to advance the ball in the opposite direction.

A line of scrimmage on the 48-yard line. The offense is on the left.

Except at the beginning of halves and after scores, the ball is always put into play by a **snap**. Offensive players line up facing defensive players at the **line of scrimmage** (the position on the field where the play begins). One offensive player, the **center**, then passes (or "snaps") the ball backwards between his legs to a teammate behind him, usually the **quarterback**.

Players can then advance the ball in two ways:

1. By running with the ball, also known as **rushing**.
2. By throwing the ball to a teammate, known as a **pass** or as **passing** the football. If the pass is thrown down-field, it is known as a

A quarterback searching for opportunity to throw a pass.

forward pass. The forward pass is a key factor distinguishing American and Canadian football from other football sports. The offense can throw the ball forward only once during a down and only from behind the line of scrimmage. However, the ball can be handed-off to another player or thrown, pitched, or tossed sideways or backwards (a lateral pass) at any time.

A down ends, and the ball becomes dead, after any of the following:

- The player with the ball is forced to the ground (a **tackle**) or has his forward progress halted by members of the other team (as determined by an **official**).
- A forward pass flies beyond the dimensions of the field (**out of bounds**) or touches the ground before it is caught. This is known as an **incomplete pass**. The ball is returned to the most recent line of scrimmage for the next down.
- The ball or the player with the ball goes out of bounds.
- A team scores.

Officials blow a whistle to notify players that the down is over.

Before each down, each team chooses a **play**, or coordinated movements and actions, that the players should follow on a down. Sometimes, downs themselves are referred to as "plays."

## Change of possession

The offense maintains possession of the ball unless one of the following things occurs:

- The team fails to get a first down— i.e., in four downs they fail to move the ball past a line 10 yards ahead of where they got their last first down. The defensive team takes over the ball at the spot where the 4th-down play ends. A change of possession in this manner is commonly called a **turnover on downs**.
- The offense scores a touchdown or field goal. The team that scored then kicks the ball to the other team in a special play called a **kickoff**.
- The offense punts the ball to the defense. A **punt** is a kick in which a player drops the ball and kicks it before it hits the ground. Punts are nearly always made on fourth down, when the offensive team does not want to risk giving up the ball to the other team at its current spot on the field (through a failed attempt to make a first down) and feels it is too far from the other team's goal post to attempt a field goal.
- A defensive player catches a forward pass. This is called an **interception**, and the player who makes the interception can run with the ball until he is tackled, forced out of bounds, or scores.
- An offensive player drops the ball (a **fumble**) and a defensive player picks it up. As with interceptions, a player recovering a fumble can run with the ball until tackled, forced out of bounds, or scoring. Passes that are thrown either backwards or parallel with the line of scrimmage (lateral passes) that are not caught do not cause the down to end as incomplete forward passes do; instead the ball is still live as if it had been fumbled. Lost fumbles and interceptions are together known as **turnovers**.
- The offensive team misses a field goal attempt. The defensive team gets the ball at the spot where the previous play began (or, in the NFL, at the spot of the kick). If the unsuccessful kick was attempted from within 20 yards

A running back being tackled when he tries to run with the ball.

A quarterback preparing to throw a pass.

Forward pass in progress, during practice.

A kicker attempts an extra point.

of the end zone, the other team gets the ball at its own 20 yard line (that is, 20 yards from the end zone). If a field goal is missed or blocked and the ball remains in the field of play, a defensive player may pick up the ball and attempt to advance it. In this last case, possession is awarded at the spot where the recovering player is ruled down.

- While in his own end zone, an offensive ball carrier is tackled, forced out of bounds, loses the ball out of bounds, or the offense commits certain fouls in the end zone. This fairly rare occurrence is called a **safety**.
- An offensive ball carrier fumbles the ball forward into the opposing end zone, and then the ball goes out of bounds. This rare occurrence leads to a **touchback**, with the ball going over to the opposing team at their 20 yard line (Note that touchbacks during non-offensive special teams plays, such as punts and kickoffs, are quite common).

## Scoring

A team scores points by the following plays:

- A **touchdown** (TD) is worth 6 points.[16] It is scored when a player runs the ball into or catches a pass in his opponent's end zone.[16] A touchdown is analogous to a try in rugby. Unlike rugby, a player does not have to touch the ball to the ground to score; a touchdown is scored any time a player has possession of the ball while any part of the ball is beyond the vertical plane created by the leading edge of the opponent's goal line stripe (the stripe itself is a part of the end zone).
  - After a touchdown, the scoring team attempts a **try** (which is also analogous to the conversion in rugby). The ball is placed at the other team's 3 yard line (the 2 yard line in the NFL). The team can attempt to kick it through the goalposts (over the crossbar and between the uprights) in the manner of a field goal for 1 point (an **extra point** or **point-after touchdown (PAT)**[17] ), or run or pass it into the end zone in the manner of a touchdown for 2 points (a **two-point conversion**). In college football, if the defense intercepts or recovers a fumble during a one or two point conversion attempt and returns it to the opposing end zone, the defensive team is awarded the two points.
- A **field goal** (FG) is worth 3 points, and it is scored by kicking the ball through the goalposts defended by the opposition.[16] Field goals may be place kicked (kicked when the ball is held vertically against the ground by a teammate) or drop kicked (extremely uncommon in the modern game due to the better accuracy of place kicks, with only two successful drop kicks in sixty-plus years in the NFL). A field goal is usually attempted on fourth down instead of a punt when the ball is close enough to the opponent's goalposts, or, when there is little or no time left to otherwise score.
- A **safety**, worth 2 points, is scored by the opposing team when the team in possession at the end of a down is responsible for the ball becoming dead behind its own goal line. For instance, a safety is scored by the defense if an offensive player is tackled, goes out of bounds, or fumbles the ball out of bounds in his own end zone.[16] Safeties are relatively rare. Note that, though even more rare, the team initially on offense during a down can score a safety if a player of the original defense gains possession of the ball in front of his own goal line and then carries the ball or fumbles it into his own end zone where it becomes dead. However, if the ball becomes dead behind the goal line of the team in possession and its opponent is responsible for the ball being there (for instance, if the defense intercepts a forward pass in its own end zone and the ball becomes dead before the ball is advanced out of the end zone) it is a touchback: no points are scored and the team last in possession keeps possession with a first down at its own 20 yard line. In the extremely rare instance that a safety is scored on a try, it is worth only 1 point.

## Kickoffs and free kicks

Each half begins with a kickoff. Teams also kick off after scoring touchdowns and field goals. The ball is kicked using a kicking tee from the team's own 35 yard line in the NFL (as of the 2011 season) and 30 yard line in college football (as of the 2007 season). The other team's kick returner tries to catch the ball and advance it as far as possible. Where he is stopped is the point where the offense will begin its **drive**, or series of offensive plays. If the kick returner catches the ball in his own end zone, he can either run with the ball, or elect for a **touchback** by kneeling in the end zone, in which case the receiving team then starts its offensive drive from its own 20 yard line. A touchback also occurs when the kick goes out-of-bounds in the end zone. (Punts and turnovers in the end zone can also result in a touchback). A kickoff that

The Florida State Seminoles (in red, at top) line up to kick off to the Virginia Tech Hokies.

goes out-of-bounds anywhere other than the end zone before being touched by the receiving team is a foul, and the ball will be placed within the hash marks of the yard line where it went out of bounds, or 30 yards from the kickoff spot, depending on which is more advantageous to the receiving team.[18] Unlike with punts, once a kickoff goes 10 yards and the ball has hit the ground, it can be recovered by the kicking team.[18] A team, especially one who is losing, can try to take advantage of this by attempting an onside kick.

After safeties, the team that gave up the points must free kick the ball to the other team from its own 20 yard line.[19]

## Penalties

Fouls (a type of rule violation) are punished with **penalties** against the offending team. Most penalties result in moving the football towards the offending team's end zone. If the penalty would move the ball more than half the distance towards the offender's end zone, the penalty becomes half the distance to the goal instead of its normal value.

Most penalties result in replaying the down. Some defensive penalties give the offense an automatic first down.[20] Conversely, some offensive penalties result in loss of a down (loss of the right to repeat the down).[20] If a penalty gives the offensive team enough yardage to gain a first down, they get a first down, as usual. The only penalty that results in points is if a team on offense commits a certain fouls, such as holding, in its own end zone, which results in a safety.

A penalty flag on the field during a game on November 16, 2008 between the San Francisco 49ers and St. Louis Rams.

If a foul occurs during a down (after the play has begun), the down is allowed to continue and an official throws a yellow penalty flag near the spot of the foul. When the down ends, the team that did not commit the foul has the option of accepting the penalty, or declining the penalty and accepting the result of the down.

## Variations

- Limited contact

  - touch football. A play ends when a defender touches the ball carrier (sometimes with two hands).
  - flag football. A play ends when a defender removes a designated token ("flag") worn by the ball carrier.
  - Wrap. A play ends when a defender wraps his arms round the ball carrier.

- Fewer players

  - nine-man football
  - eight-man football
  - six-man football

- Smaller field

  The Arena Football League is a league that plays eight-man football, but also plays indoors and on a much smaller playing surface with rule changes to encourage a much more offensive game.

- Catch and Run

  In this game, the children split into two teams and line up at opposite sides of the playing field. One side throws the ball to the other side. If the opposing team catches the ball, that player tries to run to the throwing teams touchdown without being tagged/tackled. If no one catches the ball or if the player is tagged/tackled, then that team has to throw the ball to the opposing team. This repeats until time runs out or the players decide to quit.

# Players

Most football players have highly specialized roles. At the college and NFL levels, most play only offense or only defense.

## Offense

- The **offensive line** (OL) consists of five players whose job is to protect the passer and clear the way for runners by blocking members of the defense. The lineman in the middle is the center. Outside the center are the guards, and outside them are the tackles. Except for the center, who snaps the ball to one of the backs, offensive linemen generally do not handle the ball.

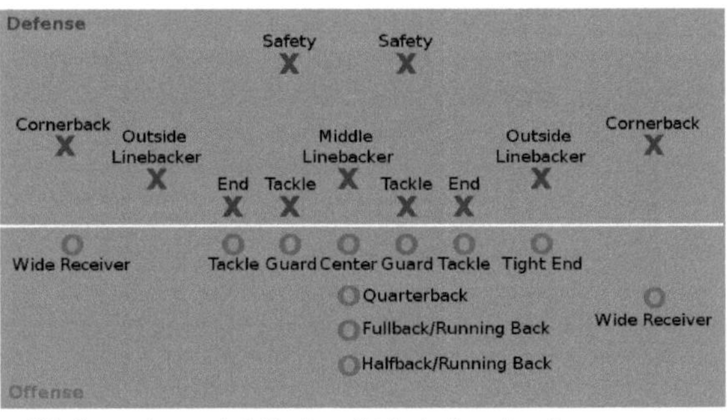


This diagram shows typical offensive and defensive formations. Because teams can change any or all of the players between plays, the number of players at certain positions may differ on a given play. Here the offense is in the Normal I-Formation while the defense is in a 4–3 Normal.

- The **quarterback** (QB) receives the snap from the center on most plays. He then hands or tosses it to a running back, throws it to a receiver or runs with it himself. The quarterback is the leader of the offense and calls the plays that are signaled to him from the sidelines.
- **Running backs** (RB) line up behind or beside the QB and specialize in running with the ball. They also block, catch passes and, on rare occasions, pass the ball to others or even receive the snap. If a team has two running backs in the game, usually one will be a **halfback** (HB) (or **tailback** (TB)), who is more likely to run with the ball, and the other will usually be a **fullback** (FB), who is more likely to block.

- **Wide receivers** (WR) line up near the sidelines. They specialize in catching passes, though they also block for running plays or downfield after another receiver makes a catch.
- **Tight ends** (TE) line up next to the offensive line. They can either play like wide receivers (catch passes) or like offensive linemen (protect the QB or create spaces for runners). Sometimes an offensive lineman takes the tight end position and is referred to as a **tackle eligible**.[21]

At least seven players must line up on the line of scrimmage on every offensive play. The other players may line up anywhere behind the line. The exact number of running backs, wide receivers and tight ends may differ on any given play. For example, if the team needs only one yard, it may use three tight ends, two running backs and no wide receivers. On the other hand, if it needs 20 yards, it may replace all of its running backs and tight ends with wide receivers.

## Defense

In contrast to members of the offense, the rules of professional football (NFL Rulebook [22]) and American college football (NCAA Rulebook [23]) do not specify starting position, movement, or coverage zones for members of the defensive team, except that they must be in the defensive zone at the start of play. The positions, movements and responsibilities of all defensive players are assigned by the team by selection of certain coverages, or patterns of placement and assignment of responsibilities. The positional roles are customary. These roles have varied over the history of American football. The following are customary defensive positions used in many coverages in modern American football.

- The **defensive line** consists of three to six players who line up immediately across from the offensive line. They try to occupy the offensive linemen in order to free up the linebackers, disrupt the backfield (behind the offensive line) of the offense, and tackle the running back if he has the ball before he can gain yardage or the quarterback before he can throw or pass the ball. They are the first line of defense.
- Behind the defensive line are the **linebackers**. They line up between the defensive line and defensive backs and may either rush the quarterback or cover potential receivers.
- The last line of defense is known as the secondary, comprising at least three players who line up as **defensive backs**, who are either **cornerbacks** or **safeties**. They cover the receivers and try to stop pass completions. They occasionally rush the quarterback.

## Special teams

The units of players who handle kicking plays are known as **special teams**. Three important special-teams players are the **punter**, who handles punts, the **placekicker** or kicker, who kicks off and attempts field goals and extra points, and the **long snapper**, who snaps the ball for extra points, field goals, and punts. Also included on special teams are the returners. These players return punts or kickoffs and try to get in good field position. These players can also score touchdowns.

## Uniform numbering

In the NFL, ranges of uniform numbers are (usually) reserved for certain positions:

- 1–19: Quarterbacks, punters and placekickers (by rule)[24]
- 20–49: Running backs and defensive backs (by rule)[25]
- 50–59: Centers and linebackers (by custom)[26]
- 60–79: Offensive guards and tackles (mandatory),[27] defensive guards and tackles (by custom)
- 10–19, 80–89: Wide receivers (by rule)[28]
- 80–89 (by rule),[29] 40-49 (optional): Tight ends
- 90–99: Defensive ends and linebackers (by custom)
- Players who switch positions in their career can keep their number if they played their prior position for at least a year and move from a position that is eligible to receive passes to another eligible position, or if he is moving from one ineligible position to another ineligible position.[30] [31]

Players wearing numbers between 50 and 79 inclusive are ineligible to receive forward passes, unless they "report as eligible" to the official.

Quarterback Shaun Carney has uniform number 5.

NCAA and high school rules specify only that offensive linemen must have numbers in the 50–79 range, but the NCAA "strongly recommends" that quarterbacks and running backs have numbers below 50 and wide receivers numbers above 79.[32] This helps officials, as it means that numbers 50 to 79 are **ineligible receivers**, or players that may not receive a forward pass (except in the rare instance when a Tackle lines up as the outermost lineman on his side of the line and the officials are notified that he will be an eligible receiver for that particular play). There are no numbering restrictions on defensive players in the NCAA, other than that a team may not have two players on the field at the same time with the same jersey number.[33]

# Basic strategy

Because the game stops after every down, giving teams a chance to call a new play, strategy plays a major role in football. Each team has a **playbook** of dozens to hundreds of plays. Ideally, each play is a scripted, strategically sound team-coordinated endeavor. Some plays are very safe; they are likely to get only a few yards. Other plays have the potential for long gains but at a greater risk of a loss of yardage or a turnover.

Generally speaking, rushing plays are less risky than passing plays. However, there are relatively safe passing plays and risky running plays. To deceive the other team, some passing plays are designed to resemble running plays and vice versa. These are referred to as play-action passes and draws, respectively. There are many trick or gadget plays, such as when a team lines up as if it intends to punt and then tries to run or pass for a first down. Such high-risk plays are a great thrill to the fans when they work. However, they can spell disaster if the opposing team realizes the deception and acts accordingly.

The defense also plans plays in response to expectations of what the offense will do. For example, a "blitz" (using linebackers or defensive backs to charge the quarterback) is often attempted when the team on defense expects a pass. A blitz makes downfield passing more difficult but exposes the defense to big gains if the offensive line stems the rush.

Many hours of preparation and strategizing, including film review by both players and coaches, go into the days between football games. This, along with the demanding physicality of football (see below), is why teams typically play at most one game per week.

## Physicality

American football is a collision sport. To stop the offense from advancing the ball, the defense must tackle the player with the ball by knocking or pulling him down. As such, defensive players must use some form of physical contact to bring the ball-carrier to the ground, within certain rules and guidelines. Tacklers cannot kick or punch the runner. They also cannot grab the face mask of the runner's helmet or lead into a tackle with their own helmet ("spearing"). Despite these and other rules regarding unnecessary roughness, most other forms of tackling are legal. Blockers and defenders trying to evade them also have wide leeway in trying to force their opponents out of the way. Quarterbacks are regularly hit by defenders coming on full speed from outside the quarterback's field of vision. This is commonly known as a blindside.

A halfback leads fellow backs through an agility drill at the Air Force Academy

To compensate for this, players must wear special protective equipment, such as a padded plastic helmet, shoulder pads, hip pads and knee pads. These protective pads were introduced decades ago and have improved ever since to help minimize lasting injury to players. An unintended consequence of all the safety equipment has resulted in increasing levels of violence in the game. Players may now hurl themselves at one another at high speeds without a significant chance of injury. The injuries that do result tend to be severe and often season or career-ending and sometimes fatal. In previous years with less padding, tackling more closely resembled tackles in Rugby football. Better helmets have allowed players to use their helmets as weapons. This form of tackling is particularly unwise, because of the great potential for brain or spinal injury. All this has caused the various leagues, especially the NFL, to implement a complicated series of penalties for various types of contact. Most recently, virtually any contact with the helmet of a defensive player on the quarterback, or any contact to the quarterback's head, is now a foul. During the late 1970s, the penalty in high school football for spearing included ejection from the game.[34]

Despite protective equipment and rule changes to emphasize safety, injuries remain very common in football. It is increasingly rare, for example, for NFL quarterbacks or running backs (who take the most direct hits) to make it through an entire season without missing some time to injury. Additionally, 28 football players died from direct football injuries in the years 2000–05 and an additional 68 died indirectly from dehydration or other examples of "non-physical" dangers, according to the National Center for Catastrophic Sport Injury Research.[35] Concussions are common, with about 41,000 suffered every year among high school players according to the Brain Injury Association of Arizona.[36] In 1981, U.S. President Ronald Reagan, who played football in high school, commented on the contact of the sport: "Football is the last thing left in civilization where men can literally fling themselves bodily at one another in combat and not be at war."[37]

Extra and optional equipment such as neck rolls, spider pads, rib protectors (referred to as "flak jackets"), and elbow pads help against injury as well, though they do not tend to be used by the majority of players due to their lack of requirement.

The danger of football, and the equipment required to reduce it, make regulation football impractical for casual play. Flag football and touch football are less violent variants of the game popular among recreational players.

## Nutrition and dehydration

Football players typically begin their season while the weather is still extremely warm and with the dangerous combination of warm weather and high humidity, dehydration is a great risk for the players.[38] The players are usually required to follow a hydration schedule. It is extremely important for players to drink enough fluids because dehydration can seriously reduce athletic performance and increase the risk of heat illnesses. Most trainers and coaches make it imperative for their players to drink fluids before they are thirsty.[38]

## Brain injury

The Concussions Committee of the NFL, co-chaired by Dr. Ira Casson, has generally denied that concussions result in permanent brain injury. However, there is some research, reported in 2009, which, using phone interviews based on the National Health Interview Survey, showed increased incidence of diagnosis of memory loss and dementia among retired professional football players; chronic traumatic encephalopathy diagnoses are increasingly common among deceased football players. Such symptoms are believed related to the effects of concussions. More rigorous research is being conducted by Dr. Casson, neurologist, for the NFL. This finding is considered significant as such injuries may potentially affect high school and college players also.[39]

# Organization in the United States

In the United States, the major forms are high school football, college football and professional football. Most American high schools field football teams. In general, high school teams play only against other teams within the same state, but there are some exceptions like nearby schools located on opposite sides of a state line.

Most of college football in the United States is governed by the National Collegiate Athletic Association (NCAA), and most colleges and universities around the country have football teams. These teams mostly play other similarly sized schools, through the NCAA's divisional system, which divides the schools into four divisions: Division I Bowl Subdivision, Division I Championship Subdivision, Division II, and Division III. Unlike the three smaller NCAA football divisions, the Division I Bowl Subdivision does not have an organized tournament to determine its national champion. Instead, teams are invited to compete in a number of post-season bowl games. In addition, the champions of six conferences in the Division I Bowl Subdivision receive automatic bids, and four other schools receive "at-large" bids, to those five bowl games under the highly lucrative Bowl Championship Series to help determine the national champion.

The highest level major professional league in the United States is the 32-team National Football League (NFL). Another professional league, the 5-team United Football League, also currently operates. Several semi-professional, women's semi-professional football, and indoor football leagues are also played across the country.

## Calendar

Amateurs playing on Thanksgiving

Although professional football has been played in every month of the year, football is traditionally an autumn sport. A season typically begins in mid-to-late August and runs through December, into January. This arrangement arose when football was in its early stages, and many of the early professional teams shared playing fields (and sometimes players) with baseball teams; the end of baseball season was accordingly set up to also serve as the start of football season. Although Thanksgiving was the traditional end of the season for most college and early professional teams, playoffs and extended seasons have pushed seasons further into the winter season. The NFL playoffs run further through January, and the Super Bowl is often played in the first week of February. As such, the sport is played in a wide range of weather conditions, from the heat and humidity of late summer, to the cold winds and snow in mid-winter. Unlike rainouts in baseball, play is only generally postponed or canceled for conditions that would pose a hazard to the safety of the fans and personnel involved, such as blizzards, hurricanes and severe thunderstorms.

Because of the amount of physical wear the game can inflict on the human body, American football seasons are typically far shorter than the comparable seasons of other popular American sports. It is very rare for football teams at any level to play more than one game in a week. For comparison, the National Hockey League and National Basketball Association traditionally play an 82-game regular season over the course of six months, plus a postseason composed of four rounds of best-of-seven series. The NFL plays 16 games over the course of four months, with a single-elimination tournament of three to four rounds.

The NFL Draft is usually held in April, in which eligible college football players are selected by NFL teams, the order of selection determined by the teams' final regular season records.

It is a long-standing tradition in the United States (though not universally observed) that high school football games are played on Friday night, college games on Saturday, and professional games on Sunday.

In the 1970s, the NFL began to schedule one game on Monday nights. Beginning in 2006, the NFL began scheduling games on Thursday nights in mid-November, aired on the NFL Network; NFL games have also been scheduled on Saturday nights toward the end of the season.

Nationally televised Thursday-night college games have become a weekly fixture on ESPN, and most nights of the week feature at least one college game, though most games are still played on the traditional Saturday.

Certain fall and winter holidays—such as the NFL's Thanksgiving Classic and numerous New Year's Day college bowl games—have traditional football games associated with them.

Despite this, there are a few professional leagues that have played in the late winter, spring, or summer, mainly to avoid competition with the established leagues. Examples include the now defunct XFL, the United States Football League, and the proposed All American Football League. Indoor football is played primarily in spring for this same reason; leagues that are based in Europe and Asia also generally play an earlier season.

At most levels of competition, college football teams hold several weeks of practices in the spring. These practices typically end with an intramural scrimmage open to the public. In certain areas, high school football teams also hold spring practices.

# Outside the United States

Outside the United States, the sport is referred to as "American football" (or a translation thereof) to differentiate it from other football codes such as association football (soccer), Rugby League, Rugby Union, Australian rules football and Gaelic football. In Australia and New Zealand the game is also known as gridiron football, or more commonly as gridiron,[1] [40] although in the United States the term *gridiron* refers only to the playing field itself.[41] The term *gridiron* has also been used in the UK to describe the game.[42] In much of the world, the term football is unambiguous and refers to association football (known commonly as 'soccer' in the United States).

## Exportation by the established U.S. leagues and organizations

In 1977, the semi-pro football teams Newton Nite Hawks and Chicago Lions completed a five-game tour in Europe.[43] These were the first professional or semi-professional American football teams to play organized games on the European continent. Games were played in Versailles, France; Lille, France; Landstuhl, Germany; Gratz Austria; and Vienna, Austria.[44]

In 1985, Bethany College head coach and future College Football Hall of Fame member Ted Kessinger brought the first American football team to play in Sweden. The Bethany "Terrible Swedes" defeated the Swedish all-star team 72–7 in Stockholm Olympic Stadium.[45]

The NFL has attempted to introduce the game to other nations and operated a developmental league, NFL Europa (also known as the World League of American Football and NFL Europe) with teams in various European cities, but this league was closed down following the 2007 season. The professional Canadian Football League and collegiate Canadian Interuniversity Sport play under the slightly different Canadian rules.

Major American leagues have also held some regular season games outside of the United States. On October 2, 2005, the Arizona Cardinals and San Francisco 49ers played the first regular season NFL game outside of the United States, in Mexico City's Estadio Azteca,[46] From 2007, the NFL has played or has plans to play at least one regular season game outside of the United States during each season. The NCAA will also play games outside of the U.S. In 2012, The United States Naval Academy will play the University of Notre Dame in Dublin, Ireland.[47]

## Leagues by country

- Australia — Gridiron Australia is the overall governing body for American football in Australia. The country is actually divided into state-level leagues instead of one national-level league by itself: ACT Gridiron (Australian Capital Territory), Gridiron NSW (New South Wales), Gridiron Queensland (Queensland), South Australian Gridiron Association (South Australia), Gridiron Victoria (Victoria), and Gridiron West (Western Australia).

- Belgium — The Belgian Football League fields 16 teams. The finalists from the playoffs determine the champion during the Belgian Bowl.

- Brazil — The Brazilian American Football League has 14 teams partitioned into north and south conferences.

- Finland — The Vaahteraliiga or the *Maple League* has eight teams. The league's name comes from the name of the championship trophy *Vaahteramalja* ("Maple Bowl"), which was donated to the newly formed association by the embassy of Canada in Finland.

- Germany — The German Football League has 12 teams partitioned into north and south conferences. The finalists from the playoffs determine the German champion during the German Bowl.

- Hungary — 18 registered teams participate in the MAFL's two-division league structure. The sport has grown significantly since 2004 and with some top Division I teams participating in the CEFL.

- India — The Elite Football League of India (EFLI) is a proposed professional league in India.[48] When play begins in late 2012, there will be eight teams, representing various cities across India with populations of one million or more. The ELFI will be India's first professional American football league, and its launch is backed by the Government of India and the Sports Authority of India.[49] All of the first season's games will be held in Pune at the Shree Shiv Chhatrapati Sports Complex.

- Ireland — The Irish American Football League consists of 14 teams. Its championship game is the Shamrock Bowl.

- Israel — Games are governed by the Israeli Football League.

- Italy — The Italian Football League was founded in 2008, taking over previous league (National Football League Italy). It has 9 teams for the 2010 season.

- Japan — The X-League is a professional league with 60 teams in four divisions, using promotion and relegation. After the post-season playoffs, the X-League champion is determined in the Japan X Bowl. There are also over 200 universities fielding teams, with the national collegiate championship determined by the Koshien Bowl. The professional and collegiate champions then face each other in the Rice Bowl to determine the national champion.

- Mexico — The ONEFA is a college league with 26 teams in 3 conferences.

- New Zealand — American Football Wellington comprises five teams located in the Wellington area.

- Norway — A rising number of teams (11 in 2010) compete in a two division league structure (division I which determines a national champion by a postseason playoff, and division 2 where newer and smaller teams are allowed to mature). Two teams (Oslo Vikings and Eidsvoll 1814s) regularly compete in either the European Football League or the EFAF Cup. Eidsvoll was the runner-up in EFAF Cup 2006.

- Poland — Games are governed by the Polish American Football League.

- Serbia — Teams in the Nacionalna Liga Srbije compete in the Serbian Bowl.

- Spain — The LNFA was founded in 1995, and currently consists of 15 clubs.

- United Kingdom — 70 amateur teams play in the BAFA Community Leagues (BAFACL) across a number of age ranges. The senior (adult) league has three levels: the Premiership, comprising six teams; Division 1, comprising 18 teams split across three regional conferences; and Division 2, comprising 23 teams split across four regional conferences. While the lower level teams have their own championship games during BritBowl Weekend, only Premier Division teams face each other in the BritBowl which is held in Worcester's Sixways Stadium. Unlike the NFL, the BAFACL season is played through the summer (April to September), with the British university season spanning the autumn and winter.

## The IFAF and the American Football World Cup

The International Federation of American Football (IFAF) is the *de facto* governing body for American football, with 45 member associations from North and South America, Europe, Asia and Oceania. The organization is headquartered in La Courneuve, France. Although the IFAF has relatively little standing in the U.S. compared to the NFL, NCAA, and the other established aforementioned bodies, these same organizations also give support to USA Football, the designated U.S. representative to the IFAF.

The IFAF also oversees the American Football World Cup, which is held every four years. Japan won the first two World Cups, held in 1999 and 2003. Team USA, which had not participated in the previous World Cups, won the title in 2007 and 2011.

A long term goal of the IFAF is for American football to be accepted by the International Olympic Committee as an Olympic sport.[50] The only time that the sport was played was at the 1932 Summer Olympics in Los Angeles, but as a demonstration sport.

## See also

- List of American football teams in the Netherlands
- List of defunct sports leagues
- Pro Football Hall of Fame
- Sports nutrition
- Sprint Football
- Steroid use in American football
- Strat-O-Matic Football
- Street football (American)

# References

## Notes

[1] O'Brien, Kerry (1999-08-04). "Gridiron comes to Australia" (http://www.abc.net.au/7.30/stories/s41614.htm). .

[2] See gridiron football for further information and sources.

[3] *See* 2006 NCAA Football Rules and Interpretations, Sec. 1, Art. 1 (http://www.ncaa.org/library/rules/2006/2006_football_rules.pdf)

[4] In the United States and Canada, the term "football" may refer to either American football or to the similar sport of Canadian football, the meaning usually being clear from the context. This article describes the American variant.

[5] "What It Was Was Football!" (http://www.library.georgetown.edu/special-collections/archives/essays/football). *Georgetown Magazine.* Georgetown University Library Special Collections. . Retrieved 2010-02-07.

[6] Bath, Richard (ed.) *The Complete Book of Rugby* (Seven Oaks Ltd, 1997 ISBN 1-86200-013-1) p77

[7] "Rutgers—The Birthplace of Intercollegiate Football" (http://www.scarletknights.com/football/history/first-game.asp). . Retrieved 2011-10-14.

[8] "Camp and His Followers: American Football 1876–1889" (http://www.footballresearch.com/articles/frpage.cfm?topic=d-to1889). *The Journey to Camp: The Origins of American Football to 1889.* Professional Football Researchers Association. . Retrieved 2007-05-16.

[9] "NFL History 1869–1910" (http://www.nfl.com/history/chronology/1869-1910). *NFL.com.* NFL Enterprises LLC. 2007. . Retrieved 2007-05-15.

[10] "NFL:America's Choice" (http://web.archive.org/web/20070808163024/http://www.coldhardfootballfacts.com/Documents/NFL_all_about_SB_1-07.pdf) (PDF). National Football League. 2007. Archived from the original (http://www.coldhardfootballfacts.com/Documents/NFL_all_about_SB_1-07.pdf) on 2007-08-08. . Retrieved 2007-08-15.

[11] "NFL Football Field Dimensions" (http://www.sportsknowhow.com/football/field-dimensions/nfl-football-field-dimensions.html). SportsKnowHow.com. 2004. . Retrieved 2009-04-05.

[12] "Coin Toss" (http://www.nfl.com/rulebook/cointoss). NFL Enterprises LLC. 2009. . Retrieved 2009-04-05.

[13] "NFL Makes Some Rule Changes" (http://ap.google.com/article/ALeqM5gHSwbXq1wnH-jmEl-jV4W8M0Vd1wD8VQBFSG0). 2008. . Retrieved 2008-04-03.

[14] "2005 Rules and Interpretations" (http://www.ncaa.org/library/rules/2005/2005_football_rules.pdf) (PDF). National Collegiate Athletic Association. 2005. . Retrieved 2008-01-09.

[15] Lawrence, Mark (2002–2005). "The Field" (http://football.calsci.com/TheRules4.html). Mark Lawrence. . Retrieved 2009-04-05.

[16] "Beginner's Guide to Football" (http://www.nfl.com/rulebook/beginnersguidetofootball). NFL Enterprises LLC. 2009. . Retrieved 2009-04-05.

[17] *2007 Official Rules of the NFL.* Triumph Books. 1 October 2007. ISBN 1699780288.

[18] "Kickoff" (http://www.nfl.com/rulebook/kickoff). NFL Enterprises LLC.. 2009. . Retrieved 2009-04-05.

[19] "Safety" (http://www.nfl.com/rulebook/safety). NFL Enterprises LLC.. 2009. . Retrieved 2009-04-05.

[20] "Penalty Summaries" (http://www.nfl.com/rulebook/penaltysummaries). NFL Enterprises LLC.. 2009. . Retrieved 2009-04-05.

[21] Member – Pro Football Hall of Fame (http://www.profootballhof.com/hof/member.jsp?player_id=158)

[22] http://www.nfl.com/rulebook

[23] http://www.ncaapublications.com/productdownloads/FR09.pdf

[24] "2011 Official Playing Rules of the National Football League" (http://static.nfl.com/static/content/public/image/rulebook/pdfs/2011_Rule_Book.pdf). National Football League. 2011. p. 23. . Rule 5, Section 1, Article 2, a

[25] "2011 Official Playing Rules of the National Football League" (http://static.nfl.com/static/content/public/image/rulebook/pdfs/2011_Rule_Book.pdf). National Football League. 2011. p. 23. . Rule 5, Section 1, Article 2, b

[26] Few centers today wear 50-range numbers, which denoted their former seldom-used status as eligible receivers. Most now wear 60–79 lineman numbers.

[27] "2011 Official Playing Rules of the National Football League" (http://static.nfl.com/static/content/public/image/rulebook/pdfs/2011_Rule_Book.pdf). National Football League. 2011. p. 23. . Rule 5, Section 1, Article 2, c

[28] "2011 Official Playing Rules of the National Football League" (http://static.nfl.com/static/content/public/image/rulebook/pdfs/2011_Rule_Book.pdf). National Football League. 2011. p. 23. . Rule 5, Section 1, Article 2, e

[29] "2011 Official Playing Rules of the National Football League" (http://static.nfl.com/static/content/public/image/rulebook/pdfs/2011_Rule_Book.pdf). National Football League. 2011. p. 23. . Rule 5, Section 1, Article 2, f

[30] Florio, Mike (March 24, 2010). "League modifies numbering system" (http://profootballtalk.nbcsports.com/2010/03/24/league-modifies-numbering-system/). *NBC Sports*. . Retrieved July 27, 2010.

[31] "2011 Official Playing Rules of the National Football League" (http://static.nfl.com/static/content/public/image/rulebook/pdfs/2011_Rule_Book.pdf). National Football League. 2011. p. 23. . Rule 5, Section 1, Article 2

[32] 2009–10 NCAA Football Rules and Interpretations (http://www.ncaapublications.com/DownloadPublication.aspx?download=FR09.pdf) Rule 1, Section 4, Article 1 Retrieved July 27, 2010

[33] 2009–10 NCAA Football Rules and Interpretations (http://www.ncaapublications.com/DownloadPublication.aspx?download=FR09.pdf) Rule 1, Section 4, Article 2, c Retrieved July 27, 2010

[34] Diehl, Jerry L., ed. *1999 and 2000 NFHS Football Handbook*. Robert F. Kanaby, NFHS Publications. p. 20.

[35] Annual Survey of Football Injury Research 1931–2005 (http://www.unc.edu/depts/nccsi/FootballInjuryData.htm), National Center for Catastrophic Sport Injury Research (http://www.unc.edu/depts/nccsi/). Updated January 18, 2006. Accessed October 31, 2006

[36] Studies Suggest 10% of Arizona High School Football Players Will Suffer a Concussion During This Coming Season (http://www.prnewswire.com/cgi-bin/stories.pl?ACCT=104&STORY=/www/story/08-23-2005/0004093186&EDATE=) PR Newswire press release from the Brain Injury Association of Arizona (http://www.biaaz.org/), August 23, 2005. Accessed October 31, 2006

[37] D'Souza, Dinesh (February 23, 1999). *Ronald Reagan: How an Ordinary Man Became an Extraordinary Leader*. Free Press. p. 40. ISBN 0684848236.

[38] Hemmelgran, Melinda. "Nutrient Needs of Young Athletes." The Elementary School Journal: Sports and Physical Education 91 (1991): 445–56.

[39] "Dementia Risk Seen in Players in N.F.L. Study" (http://www.nytimes.com/2009/09/30/sports/football/30dementia.html) article by Alan Schwarz in *The New York Times* September 29, 2009

[40] Louis S. Leland (1984) personal Kiwi-Yankee dictionary (http://books.google.com/books?id=FqlEe5d4yJ8C&pg=PA39&dq=gridiron+new+zealand&hl=en&ei=SbYPTPO1McalOLessPwK&sa=X&oi=book_result&ct=result&resnum=7&ved=0CEwQ6AEwBg#v=onepage&q=gridiron new zealand&f=false) Pelican Publishing Company, 1984

[41] "gridiron." The American Heritage Dictionary of the English Language, Fourth Edition. Houghton Mifflin Company, 2004. 01 October 2007. (http://dictionary.reference.com/browse/gridiron).

[42] Kelso, Paul (2007-10-27). Gridiron: on its way to Wembley. (http://www.guardian.co.uk/uk/2007/oct/27/uknews4.mainsection1) (London) The Guardian, 27 October 2007.

[43] "Europe to get a look at American football" (http://news.google.com/newspapers?id=c_opAAAAIBAJ&sjid=lV0DAAAAIBAJ&pg=6864,2898012&dq=chicago-lions+europe+football+newton&hl=en). St. Petersburg Times. May 27, 1977. . Retrieved January 21, 2011.

[44] "Pro grid coaches still convinced Europe fertile" (http://news.google.com/newspapers?id=tL0qAAAAIBAJ&sjid=fWcEAAAAIBAJ&pg=6819,3673442&dq=chicago-lions+europe+football+newton&hl=en). Sarasota Herald-Tribune. June 24, 1977. . Retrieved January 21, 2011.

[45] Nastrom, Stephaan (June 20, 1985). "Sweden's First Shot at Football a Success Despite 72–7 Defeat" (http://news.google.com/newspapers?nid=861&dat=19850620&id=4XAoAAAAIBAJ&sjid=R4UDAAAAIBAJ&pg=6733,4809175). *The Victoria Advocate*. . Retrieved January 21, 2011.

[46] Young, Eric (2005-03-16). "S.F. 49ers, Arizona Cardinals to kick off in Mexico" (http://www.bizjournals.com/sanfrancisco/stories/2005/03/14/daily27.html). .

[47] "Notre Dame And Navy Extend Series 10 More Years :: Irish and Midshipmen will meet at least until 2016, with the 2012 meeting set to be in Dublin, Ireland" (http://und.cstv.com/sports/m-footbl/spec-rel/111005aac.html). Und.cstv.com. . Retrieved 2008-11-24.

[48] http://www.efli.com/letter-from-the-founder.php

[49] http://sports.yahoo.com/nfl/news?slug=ycn-8919377

[50] Mike Florio (2010-02-24). "Football not truly global until it's in Olympics" (http://nbcsports.msnbc.com/id/35563184/ns/sports-nfl/). MSNBC.com. . Retrieved 2010-02-27.

## Bibliography

- "Digest of Rules" (http://www.nfl.com/rulebook/digestofrules). National Football League. Retrieved 2007-10-31.
- "History and the basics" (http://www.nfl.com/history). National Football League. Retrieved 2005-12-28.
- "Playing with the Percentages When Trailing by Two Touchdowns" (http://web.archive.org/web/20051127032944/http://www.thesportjournal.org/2005Journal/Vol8-No4/starkey.asp). Montana State University. Archived from the original (http://www.thesportjournal.org/2005Journal/Vol8-No4/starkey.asp) on 2005-11-27. Retrieved 2005-12-24.

## Further reading

- *Sports Illustrated* magazine dated December 4, 2005; "Football America", a series of articles attesting to the pervasive popularity of American football in the United States at all levels.

## External links

- Movie of 1903 football game between the University of Chicago and the University of Michigan (http://memory.loc.gov/cgi-bin/query/S?ammem/papr:@FILREQ(@field(TITLE+@od1(Chicago-Michigan+football+game++))+@FIELD(COLLID+workleis)))
- National Football League Official Signals (http://www.nfl.com/rulebook/signals).
- Brief explanation of the sport by the BBC aimed at a non-american audience (http://news.bbc.co.uk/sport1/hi/other_sports/american_football/3192002.stm)
- American Football (http://www.dmoz.org/Sports/Football/American//) at the Open Directory Project
- American Youth Football (http://www.americanyouthfootball.com/)

# Wayne_Bolt

<table>
<tr><td colspan="2" align="center">Wayne Bolt</td></tr>
<tr><td>Sport(s)</td><td>Football</td></tr>
<tr><td colspan="2" align="center">Current position</td></tr>
<tr><td>Title</td><td>Director of Football Relations</td></tr>
<tr><td>Team</td><td>Auburn University</td></tr>
<tr><td>Conference</td><td>SEC</td></tr>
<tr><td colspan="2" align="center">Biographical details</td></tr>
<tr><td>Born</td><td>unknown</td></tr>
<tr><td>Place of birth</td><td>Augusta, Georgia</td></tr>
<tr><td colspan="2" align="center">Playing career</td></tr>
<tr><td>1976-1980</td><td>East Carolina</td></tr>
<tr><td>Position(s)</td><td>Offensive line</td></tr>
<tr><td colspan="2" align="center">Coaching career (HC unless noted)</td></tr>
<tr><td>1980<br>1981-1985<br>1986-1989<br>1990<br>1991-1996<br>1997-2002<br>2002-2005<br>2006-2008<br>2009-present</td><td>Wyoming (Assistant)<br>Auburn (TE)<br>Clemson (TE)<br>Auburn (A)<br>Troy (OL)<br>Troy (DC/AHC)<br>UAB (DC)<br>Iowa State (DC)<br>Auburn (A)</td></tr>
</table>

**Wayne Bolt** (born in Augusta, Georgia) is an American football coach, currently serving as the Director of Football Relations for Auburn University. Bolt took the job in 2009 upon the urging of head coach Gene Chizik. Prior to joining the Auburn staff, Bolt had previously served as defensive coordinator to Chizik at Iowa State University. Bolt also had previous connections to Auburn, having served as an assistant at the school under head coach Pat Dye.

Bolt has coached 10 teams that made bowl games at UAB, Troy, Clemson, and Auburn. During his time as a position coach, he tutored over 20 players who went on to play in the NFL.

## Career

Bolt is a 1974 graduate of the Academy of Richmond County and played college football at East Carolina University, where he was an All-American offensive lineman. Upon graduation, he joined the staff as an assistant coach. When head coach Pat Dye was hired away from ECU, Bolt joined Dye's staff at Wyoming. When Dye took the head coaching position at Auburn in 1981, Bolt again followed his mentor to the Plains as the tight end coach. Bolt left Auburn to join Danny Ford's staff at Clemson as tight-ends coach in 1986, but again returned to Auburn in 1990.

In 1991, Bolt joined the staff at Troy University, working under fellow former Auburn assistant Larry Blakeny as the offensive line coach. In 1997, he was promoted to defensive coordinator and assistant head coach for the Trojans. Bolt's defenses were extremely successful at Troy. In 2002, in only the second season competing at the NCAA I-A level, the Trojans produced one of the nation's top defenses under Bolt's guidance. The Trojans ranked 4th nationally in total defense, yielding only 276.8 yards per game. The unit ranked 13th in rushing defense (105.3 yards per game), 30th in scoring defense (21.0 ppg) and 33rd in pass efficiency defense (112.42).

Following his success at Troy, Bolt left the school in January 2002 to head the defense at the University of Alabama at Birmingham. He served on Watson Brown's staff until being fired from his position with the Blazers following the 2005 season.[1]

New head coach Gene Chizik hired Bolt to join his staff at Iowa State University in December 2006. When Chizik left the Cyclones to take the vacant head coaching position at Auburn, he brought Bolt along in an administrative capacity. He was named the Director of Football Operations for the Tigers in March 2009.[2]

## External links

- Official Iowa State bio [3]
- Official UAB bio [4]

## References

[1]  http://uabsports.cstv.com/sports/m-footbl/spec-rel/121505aac.html

[2]  http://auburntigers.cstv.com/sports/m-footbl/spec-rel/032409aaa.html

[3]  http://www.cyclones.com/ViewArticle.dbml?DB_OEM_ID=10700&ATCLID=715115

[4]  http://uabsports.cstv.com/sports/m-footbl/mtt/bolt_wayne00.html

# Auburn_University

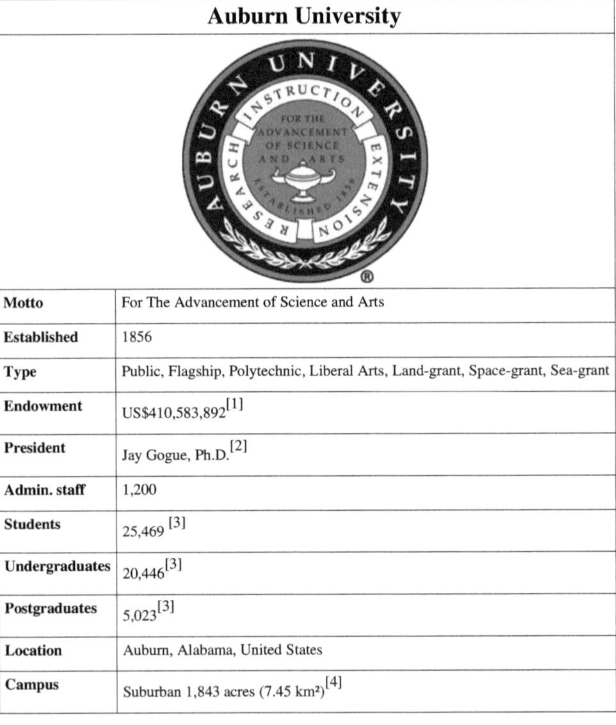

| Auburn University | |
|---|---|
| **Motto** | For The Advancement of Science and Arts |
| **Established** | 1856 |
| **Type** | Public, Flagship, Polytechnic, Liberal Arts, Land-grant, Space-grant, Sea-grant |
| **Endowment** | US$410,583,892[1] |
| **President** | Jay Gogue, Ph.D.[2] |
| **Admin. staff** | 1,200 |
| **Students** | 25,469 [3] |
| **Undergraduates** | 20,446[3] |
| **Postgraduates** | 5,023[3] |
| **Location** | Auburn, Alabama, United States |
| **Campus** | Suburban 1,843 acres (7.45 km²)[4] |

| Former names | East Alabama Male College (1856–1872) |
| --- | --- |
| | Agricultural and Mechanical College of Alabama (1872–1899) |
| | Alabama Polytechnic Institute (1899–1960) |
| Athletics | Auburn Tigers |
| Colors | Burnt Orange (PMS 158) and Navy Blue (PMS 289) [5] |
| Mascot | Aubie the Tiger (costumed) |
| Website | auburn.edu [6] |

**Auburn University** (**AU** or **Auburn**) is a public university located in Auburn, Alabama, United States. With more than 25,000 students and 1,200 faculty members, it is one of the largest universities in the state.[7] Auburn was chartered on February 7, 1856, as the **East Alabama Male College**,[8] a private liberal arts school affiliated with the Methodist Episcopal Church, South. In 1872, the college became the state's first public land-grant university under the Morrill Act and was renamed the **Agricultural and Mechanical College of Alabama**.[9] In 1892, the college became the first four-year coeducational school in the state. The curriculum at the university originally focused on arts and agriculture. This trend changed under the guidance of Dr. William Leroy Broun, who taught classics and sciences and believed both disciplines were important in the overall growth of the university and the individual. The college was renamed the **Alabama Polytechnic Institute** (API) in 1899, largely because of Dr. Broun's influence.[9] The college continued expanding, and in 1960 its name was officially changed to Auburn University to acknowledge the varied academic programs and larger curriculum of a major university. It had been popularly known as "Auburn" for many years.[10] Auburn is among the few American universities designated as a land-grant, sea-grant, and space-grant research center.

# History

"Old Main," the first building on Auburn's campus, was destroyed by fire in 1887

Auburn University was chartered by the Alabama Legislature as the East Alabama Male College on February 7, 1856, coming under the guidance of the Methodist Church in 1859.[11] The first president of the institution was Reverend William J. Sasnett, and the school opened its doors in 1859 to a student body of eighty and a faculty of ten. The early history of Auburn is inextricably linked with the Civil War and the Reconstruction-era South. Classes were held in "Old Main" until the college was closed due to the Civil War, when most of the students and faculty left to enlist. The campus was used as a training ground for the Confederate Army, and "Old Main" served as a hospital for Confederate wounded.

To commemorate Auburn's contribution to the Civil War, a cannon lathe used for the manufacture of cannons for the Confederate Army and recovered from Selma, Alabama, was presented to Auburn in 1952 by brothers of Delta Chapter of the Alpha Phi Omega fraternity.[12] It sits today on the lawn next to Samford Hall.

## Post-Civil War

The school was reopened in 1866 following the end of the Civil War and has been open ever since. In 1872, control of the institution was transferred from the Methodist Church to the State of Alabama for financial reasons. Alabama placed the school under the provisions of the Morrill Act as a land-grant institution, the first in the South to be established separate from the state university. This act provided for 240,000 acres (971 km²) of Federal land to be sold in order to provide funds for an agricultural and mechanical school. As a result, in 1872 the school was renamed to the Agricultural and Mechanical College of Alabama.

Under the provisions of this act, land-grant institutions were also supposed to teach military tactics and train officers for the United States military. In the late 19th century, most students at the Agricultural and Mechanical College of Alabama were enrolled in the cadet program, learning military tactics and training to become future officers. Each county in the state was allowed to nominate two cadets to attend the college free of charge.

In 1892, two historic events occurred: women were first admitted to the Agricultural and Mechanical College of Alabama, and football was first played as a school sport. Eventually, football replaced polo as the main sport on campus. In 1899, the school name was again changed, this time to Alabama Polytechnic Institute.

Samford Hall in the 1890s

API Cadets drill on Ross Square in 1918.

On October 1, 1918, nearly all of Alabama Polytechnic Institute's able-bodied male students 18 or older voluntarily joined the United States Army for short-lived military careers on campus. The student-soldiers numbered 878, according to API President Charles Thach, and formed the academic section of the Student Army Training Corps. The vocational section was composed of enlisted men sent to Auburn for training in radio and mechanics. The students received honorable discharges two months later following the Armistice that ended World War I.

API struggled through the Great Depression, having scrapped an extensive expansion program by then-President Bradford Knapp. Faculty salaries were cut drastically, and enrollment decreased along with State appropriations to the college. By the end of the 1930s, Auburn had essentially recovered, but then faced new conditions caused by World War II.

As war approached in 1940, there was a great shortage of engineers and scientists needed for the defense industries. The U.S. Office of Education asked all American engineering schools to join in a 'crash' program to produce what was often called 'instant engineers.' API became an early participant in an activity that eventually became Engineering, Science, and Management War Training (ESMWT). Fully funded by the government and coordinated by Auburn's Dean of Engineering, college-level courses were given in concentrated, mainly evening classes at sites across Alabama. Taken by thousands of adults – including many women – these courses were highly beneficial in filling the wartime ranks of civilian engineers, chemists, and other technical professionals. The ESMWT also benefited API by providing employment for faculty members when the student body was significantly diminished by the draft and patriotic volunteers.

During the war, API also trained U.S. military personnel on campus; between 1941 and 1945, Auburn produced over 32,000 troops for the war effort. Following the end of World War II, API, like many colleges around the country, experienced a period of massive growth caused by returning military personnel taking advantage of their GI Bill offer of free education. In the five-year period following the end of the war, enrollment at API more than doubled.

Modern logo

## Name change: Auburn

Recognizing the school had moved beyond its agricultural and mechanical roots, it was granted university status by the Alabama Legislature in 1960 and officially renamed Auburn University, a name that better expressed the varied academic programs and expanded curriculum that the school had been offering for years. However, it had been popularly called "Auburn" for many years even before the official name change.

Like most universities in the American South, Auburn was racially segregated prior to 1963, with only white students being admitted. Integration went smoothly at Auburn, with the first African-American student being admitted in 1964, and the first doctoral degree being granted to an African-American in 1967.

Today, Auburn has grown since its founding to have an on-campus enrollment of over 25,000 students and a faculty of almost 1,200 at the main campus in Auburn.[13] There are also over 6,000 students at the Auburn University Montgomery satellite campus established in 1967.

# Academics

Samford Hall, located on College Street in
Auburn, houses the University's administration.

| University rankings (overall) | |
|---|---|
| **National** | |
| *Forbes*[14] | 116 |
| *U.S. News & World Report*[15] | 82 |
| *Washington Monthly*[16] | 93 |
| **Global** | |
| *ARWU*[17] | 401–500 |
| *Times*[18] | 351–400 |

| Auburn rankings | |
|---|---|
| *USNWR* National University[19] | 82 |
| *USNWR* National Top Publics[20] | 36 |
| *USNWR* Business schools[21] | 63 |
| *USNWR* Education[22] | 71 |
| *USNWR* Engineering[23] | 70 |
| *USNWR* Veterinary Medicine[24] | 15 |
| *USNWR* Pharmacy[25] | 24 |
| *USNWR* Public Affairs[26] | 57 |

| | |
|---|---|
| *USNWR* **RehabilitationCounseling**[27] | 17 |
| *USNWR* **Audiology**[26] | 46 |
| *USNWR* **psychology**[26] | 103 |
| *USNWR* **Clinical psychology**[26] | 83 |
| *USNWR* **Computer Science**[26] | 91 |
| *USNWR* **Biology**[26] | 100 |
| *USNWR* **English**[26] | 94 |

Auburn has traditionally been rated highly by academic ranking services, and has been listed as one of the top 50 public universities for 17 consecutive years.[28] The 2011 edition of *U.S. News and World Reports* ranks Auburn as the 85th university in the nation among public and private schools and 38th among public universities.[29] Auburn was the only college or university in Alabama included in the inaugural edition (1981) of the widely respected *Peterson's Guides to America's 296 Most Competitive Colleges.*

Auburn is a charter member of the Southeastern Conference (SEC), which is currently composed of 11 of the largest Southern public universities in the US and one private university, Vanderbilt. Among the other 10 peer public universities; the University of Florida, the University of Georgia, and the University of Alabama are ranked ahead of Auburn in the 2011 edition of *U.S. News & World Report.*[30] This high ranking and reputation for academic quality is in spite of the fact that Auburn's $378.6 million endowment being currently the second smallest of the 12 SEC universities.[31] An attempt to increase the endowment by $500 million began in 2005 with the "It Begins at Auburn" campaign. As of August 2006, the campaign had raised $523 million, making it the largest campaign in university history.[32]

The university currently consists of thirteen schools and colleges. Programs in architecture, pharmacy, veterinary science, engineering, forestry, and business have been ranked among the best in the country.

The journal *DesignIntelligence* in its 2009 edition of "America's Best Architecture and Design Schools" ranked Auburn's undergraduate Architecture program No. 12 and Industrial Design program No. 7 nationally, and the deans of Architecture schools ranked the Architecture program the second most admired. In addition, Auburn's graduate Landscape Architecture program was ranked No. 14 nationally and Industrial Design program 5th.

The Department of Foundations, Leadership and Technology of the College of Education was ranked 7th in the nation by Academic Analytics in 2008.

Auburn University's College of Architecture pioneered the joining of architecture and interior design curriculum with the nation's first interior architecture degree program. The Dual Degree Architecture & Interior Architecture degree was the first in the nation as well. Auburn University's College of Architecture, Design, and Construction also pioneered the nations first Design Build Master's Degree program, hence capitalizing on The College of Architecture, Design and Construction's "Building Science" program with Auburn's "Rural Studio" program where Architectural students build highly creative and ingenious homes for some of the poorest regions of Alabama. These homes and efforts have been publicized by People Magazine, Time, featured on Oprah Winfrey, numerous Architectural and Construction periodicals as well. Of critical mention here is the School's Rural Studio program, founded by the late Samuel Mockbee.

The Ginn College of Engineering has a 134-year tradition of engineering education, consistently ranking in the nation's top 20 engineering programs in terms of numbers of engineers graduating annually. The college has a combined enrollment of close to 4,000. Auburn's College of Engineering offers majors in civil, mechanical, electrical, industrial and systems engineering, polymer and fiber engineering, aerospace, agricultural, bio-systems, materials, chemical engineering, computer science, and software engineering, and—more recently—began a program in wireless engineering after receiving a donation from alumnus Samuel L. Ginn. In 2001, Ginn, a noted US pioneer in wireless communication, made a $25 million gift to the college and announced plans to spearhead an additional $150 million in support. This gave Auburn the first Bachelor of Wireless Engineering degree program in the United States. Auburn University was the first university in the Southeast to offer the bachelor of software engineering degree and the master of software engineering degree.

Ross Chemical Laboratory

Auburn has historically placed much of its emphasis on the education of engineers at the undergraduate level, and in recent years has been ranked as high as the 10th largest undergraduate engineering program in the US in terms of the number of undergraduate degrees awarded on annual basis. The Ginn College of Engineering is now focused on growing the graduate programs, and recent rankings demonstrate the increasing profile of graduate engineering education at Auburn. The Ginn College of Engineering was recently ranked 60th nationally overall and 35th among public universities that offer doctoral programs in engineering by *U.S. News and World Report*. Last year, the College ranked 67th among all engineering programs and 40th among such programs at public universities. "America's Best Graduate Schools 2006" ranks the Ginn College of Engineering's graduate program in the Top 100 graduate engineering programs in the US. Auburn's Industrial and Systems Engineering, Civil Engineering, Chemical Engineering, and Mechanical Engineering were all ranked in the top 100.

Auburn also boasts strong programs in veterinary medicine, mathematics, science, agriculture, and journalism. The university's core curriculum has likewise been recognized as one of the best in the nation.[citation required]

Auburn's Economics Department (formerly in the College of Business, now in the College of Liberal Arts) was ranked 123rd in the world in 1999 by the *Journal of Applied Econometrics*. Auburn was rated ahead of such international powerhouses as INSEAD in France (141st) and the London Business School (146th). Auburn's MBA Program in the College of Business has annually been ranked by *U.S. News and World Report magazine* in the top ten percent of the nation's more than 750 MBA Programs. The Ludwig von Mises Institute offices were once located in the business department of Auburn University, and the LvMI continues to work with the university on many levels.[33]

Nationally recognized ROTC programs are available in three branches of service: Air Force, Army, and Navy/Marine Corps, the latter being the only one of its kind in Alabama. Each of these three ROTC units is ranked among the top ten in the nation. Over 100 officers that attended Auburn have reached flag rank (general or admiral), including one, Carl Epting Mundy Jr., who served as Commandant of the US Marine Corps. Auburn is one of only seven universities in the Nuclear Enlisted Commissioning Program, and has historically been one of the top ROTC producers of Navy nuclear submarine officers.

In addition to the many outstanding ROTC graduates commissioned through Auburn, two masters degree alumni from Auburn, four-star generals Hugh Shelton and Richard Myers, served as Chairman of the Joint Chiefs of Staff in the last decade. Both officers received their commissions elsewhere, and attended Auburn for an M.S. (Shelton) and M.B.A. (Myers).

Auburn has graduated six astronauts (including T.K. Mattingly of Apollo 13 fame) and one current and one former director of the Kennedy Space Center. 1972 Auburn Mechanical Engineering graduate Jim Kennedy, currently director of NASA's Kennedy Space Center, was previously deputy director of NASA's Marshall Space Flight Center (MSFC). Several hundred Auburn graduates, primarily engineers and scientists, currently work directly for NASA or NASA contractors. Hundreds of Auburn engineers worked for NASA at MSFC during the peak years of the "space race" in the 1960s, when the Saturn and Apollo moon programs were in full development.

Jordan-Hare Stadium, Beard-Eaves-Memorial Coliseum, and Samford Stadium-Hitchcock Field at Plainsman Park on the Auburn University campus

Auburn University owns and operates the 423-acre (1.71 km$^2$) Auburn-Opelika Robert G. Pitts Airport, providing flight education and fuel, maintenance, and airplane storage. The Auburn University Aviation Department is fully certified by the FAA as an Air Agency with examining authority for private, commercial, instrument, and multiengine courses. The College of Business's Department of Aviation Management and Supply Chain Management is the only program in the country to hold dual accreditation by both the Association to Advance Collegiate Schools of Business (AACSB) and the Aviation Accreditation Board International (AABI).[34] Created over 65 years ago, Auburn's flight program is also the second oldest university flight program in the United States.[34]

Auburn University has been recognized as having some of the best agriculture, fisheries, forestry, and poultry science programs in the South. The Old Rotation on campus is the oldest continuous agricultural experiment in the Southeast, and third oldest in the United States, dating from 1896. In addition, the work of Dr. David Bransby on the use of switchgrass as a biofuel was the source of its mention in the 2006 State of the Union Address.

The university recently began a Master of Real Estate Development program.[35] This is one of the few in the Southeast, with primary competition with the University of Central Florida, University of Florida, University of South Florida, and Clemson University. The program has filled a void of professional real estate education in Alabama.

*Modern Healthcare* ranked Auburn University's Physicians Executive M.B.A. (PEMBA) program in the College of Business ninth in the nation among all degree programs for physician executives, according to the Journal's May 2006 issue. Among M.B.A. programs tailored specifically for physicians, AU's program is ranked second.

## Schools and Departments

*Date indicated is year of founding*

- College of Agriculture, 1872

    - Agricultural Economics and Rural Sociology
    - Agronomy and Soils
    - Animal Sciences
    - Biosystems Engineering
    - Entomology and Plant Pathology
    - Fisheries and Allied Aquacultures
    - Horticulture
    - Poultry Science

- College of Architecture, Design and Construction, 1907

- College of Business, 1967

    - School of Accountancy
    - Aviation and Supply Chain Management
    - Finance
    - Management
    - Marketing

- College of Education, 1915

    - Curriculum and Teaching
    - Educational Foundations, Leadership and Technology
    - Kinesiology
    - Special Education, Rehabilitation, Counseling/School Psychology

- Samuel Ginn College of Engineering, 1872

- School of Forestry and Wildlife Sciences, 1984

- College of Human Sciences, 1916

    - Consumer Affairs
    - Human Development and Family Studies
    - Nutrition, Dietetics, and Hospitality Management

- College of Liberal Arts, 1986

- School of Nursing, 1979

- James Harrison School of Pharmacy, 1885

    - Pharmacal Sciences
    - Pharmacy Practice
    - Pharmacy Care Systems

- College of Sciences and Mathematics, 1986

    - Biological Sciences
    - Chemistry and Biochemistry
    - Geology and Geography
    - Mathematics and Statistics
    - Physics

- College of Veterinary Medicine, 1907

    - Anatomy, Physiology and Pharmacology

  - Clinical Sciences
  - Pathobiology
- Graduate School, 1872

## Campus arrangement

The Auburn campus is primarily arranged in a grid-like pattern with several distinct building groups. The northern section of the central campus (bounded by Magnolia Ave. and Thach Ave.) contains most of the College of Engineering buildings, the Lowder business building, and the older administration buildings. The middle section of the central campus (bounded by Thach Ave. and Roosevelt Dr.) contains the College of Liberal Arts (except fine arts) and the College of Education, mostly within Haley Center. The southern section of the central campus (bounded by Roosevelt Dr. and Samford Ave.) contains the most of the buildings related to the College of Science and Mathematics, as well as fine arts buildings.

Several erratic building spurts, beginning in the 1950s, have resulted in some exceptions to the subject clusters as described above. Growing interaction issues between pedestrians and vehicles led to the closure of a significant portion of Thach Avenue to vehicular traffic in 2004. A similarly sized portion of Roosevelt Drive was also closed to vehicles in 2005. In an effort to make a more appealing walkway, these two sections have been converted from asphalt to concrete. The general movement towards a pedestrian only campus is ongoing, but is often limited by the requirements for emergency and maintenance vehicular access.

The current period of ongoing construction began around the year 2000. All recently constructed buildings have used a more traditional architectural style that is similar to the style of Samford Hall, Mary Martin Hall, and the Quad dorms. The Science Center complex was completed in 2005. This complex contains chemistry labs, traditional classrooms, and a large lecture hall. A new medical clinic opened behind the Hill dorm area. Taking the place of the old medical clinic and a few other older buildings, is the Shelby Center for Engineering Technology. Phase I of the Shelby Center opened in the Spring of 2008, with regular classes being held starting with the Summer 2008 term. A new Student Center opened in 2008. [36] [37]

## Student life

### Housing

Auburn's initial Campus Master plan was designed by Frederick Law Olmsted. For most of the early history of Auburn, boarding houses and barracks made up most of the student housing. Even into the 1970s, boarding houses were still available in the community. It wasn't until the great depression that Auburn began to construct the first buildings on campus that were "dorms" in the modern sense of the word. As the university gradually shifted away from agricultural and military instruction to more of an academic institution, more and more dorms began to replace the barracks and boarding houses.

Auburn's first dorms were hardly luxurious. Magnolia Dormitory, built in the 1950s and demolished in 1987, was once used by the state of Alabama in its defense against a lawsuit brought by state prison inmates. The inmates claimed that housing two men in a cell of particularly small dimensions constituted 'cruel and unusual punishment.' The state argued in court that students at Auburn actually paid to live in even smaller living spaces—at Magnolia Dorm. The inmates lost the case. Its "twin", Noble Hall, used exclusively for women, was demolished in 2005 and was condemned during at least the final year in which it was inhabited.

In the last thirty years, the city of Auburn has experienced a rapid growth in the number of apartment complexes constructed. Most Auburn students today live off-campus in the apartment complexes and condos, which surround the immediate area around the university. Only 19 percent of all undergraduate students at Auburn live on campus.[38]

Auburn's on-campus student housing consists of four complexes located at various locations over campus — "The Quad", "The Village", "The Hill", and "The Extension". "The Quad" is the oldest of the four, dating to the Great Depression projects begun by the Works Progress Administration and located in Central Campus. Made up of ten buildings, the Quad houses undergraduate students. Eight of the buildings are coed by floor, the remaining two are female-only.

"The Hill" is made up of 12 buildings and is located in South Campus. The Hill houses mostly undergraduates. There are two high-rise, 6-story dormitories (Boyd and Sasnett), and all dorms are coed (but have gender-separated floors) with the exception of Leischuck and Hall M (home of the M Gym, the campus's only exercise center that charges extra fees), which are female only. All of the Hill dormitories were used to house sororities until 2009. The sororities are now housed in the newly completed village. "The Hill" is the cheapest on-campus housing at Auburn due to the substandard nature of Terrell dining hall, located in the middle of "The Hill" dormitories.

"The Extension" is a block of six buildings (labeled A, B, C, D, E, and F), each consisting of two-bedroom apartments, housing undergraduates. The extension closed in 2009, but has been re-opened for the 2010–2011 school year.

"The Village," formerly known as married student housing, recently housed a variety of students, to include undergraduates, graduates, and married students. In May 2006, this housing complex was closed to students and was demolished during the summer and early fall of 2006; however, in 2009 it was rebuilt into 8 4-story buildings to accommodate 1,700 residents. This area now houses sororities and undergrads.[39]

## Greek life

Greek associated students make up roughly 24 percent of undergraduate men and 34 percent of women at Auburn.

Male Greeks in Auburn are roughly divided into two separate areas: Old Row and New Row. "Old Row" traditionally was made up of the fraternities whose houses were located along Magnolia Avenue on the north side of campus. "New Row" is made up of fraternities whose houses were located along Lem Morrison Drive southwest of campus. However, being an "Old Row" or "New Row" fraternity doesn't really depend on where the house is located but on the age of the fraternity. Therefore, there are some "Old Row" fraternities with houses on "New Row" Lem Morrison Drive because they moved there. Today's "Old Row" on and around Magnolia Avenue was once the "New Row," as the first generation of fraternity houses at Auburn were on or near College Street. Most of these houses were demolished by the end of the 1970s, and only two fraternity houses remain on College Street today.

There are seventeen social sororities represented at Auburn University. Sorority recruitment is a week-long process held by the Panhellenic Council in August every year. Sororities are located not in individual houses like Auburn fraternities, but in the designated dorms located in The Village. This has the unintended side effect of keeping dues for these sororities among the lowest in the nation. Each dorm has a sorority "chapter" room within it for the sorority designated to that dorm.

# Athletics

Auburn University's sports teams are known as the Tigers, and they participate in Division I-A of the NCAA and in the Western Division of the 14-member Southeastern Conference (SEC). Auburn has won a total of 18 intercollegiate national championships (including 16 NCAA Championships), which includes 2 football (1957, 2010), 8 men's swimming and diving (1997, 1999, 2003, 2004, 2005, 2006, 2007, 2009), 5 women's swimming and diving (2002, 2003, 2004, 2006, 2007), 2 equestrian (2008, 2011), and 1 women's outdoor track and field (2006) titles. Auburn has also won a total of 70 Southeastern Conference championships, including 51 men's titles and 19 women's titles. Auburn's colors of burnt orange and navy blue were chosen by Dr. George Petrie, Auburn's first football coach, based on those of his alma mater, the University of Virginia.

Aubie, The Auburn University Tiger Mascot

## Football

Auburn's football program is currently coached by Gene Chizik. Past coaches include George Petrie, John Heisman, Mike Donahue, Ralph "Shug" Jordan, Pat Dye, Terry Bowden and Tommy Tuberville.

Auburn played its first game in 1892 against the University of Georgia at Piedmont Park in Atlanta, Georgia starting what is currently the oldest college football rivalry in the Deep South. The Tigers' first bowl

Tiger statue outside Jordan-Hare Stadium

appearance was in 1937 in the sixth Bacardi Bowl played in Havana, Cuba. AU football has won seven SEC Conference Championships, and since the division of the conference in 1992, seven western division championships and four trips to the SEC Championship game. Auburn plays arch-rival Alabama each year in a game known as the Iron Bowl.

In 1957, Auburn was coached by "Shug" Jordan to a 10–0 record and was awarded the AP National Championship. Ohio State University was first in the UPI coaches' poll. Auburn was ineligible for a bowl game, however, having been placed on probation by the Southeastern Conference.

Three Auburn players, Pat Sullivan in 1971, Bo Jackson in 1985, and Cam Newton in 2010 have won the Heisman Trophy. The Trophy's namesake, John Heisman, coached at Auburn from 1895 until 1899. Auburn is the only school where Heisman coached (among others, Georgia Tech and Clemson) that has produced a Heisman Trophy winner. Auburn's Jordan-Hare Stadium has a capacity of 87,451 ranking as the ninth-largest on-campus stadium in the NCAA as of September 2006.

Jordan-Hare Stadium (2005)

Auburn went 11–0 under Terry Bowden in 1993, but was on probation and not allowed to play in the SEC Championship game. Auburn completed the 2004 football season with a 13–0 record winning the SEC championship, the school's first conference title since 1989 and the first outright title since 1987. The 2004 team was led by quarterback Jason Campbell, running backs Carnell Williams and Ronnie Brown, and cornerback Carlos Rogers, all subsequently drafted in the first round of the 2005 NFL Draft. The team's new offensive coordinator, Al Borges, led the team to use the west coast style offense which maximized the use of both star running backs. However, the Tigers were ranked behind two other undefeated teams, Southern California and Oklahoma, that played in the BCS championship game.

Prior to the 2008 season, Tony Franklin was hired as offensive coordinator to put Auburn into the spread offense. He was fired, however, following the sixth game of the season that ended in a loss to Vanderbilt. Tommy Tuberville then resigned as head coach after the season. On December 13, 2008, it was reported that Gene Chizik had been hired as Auburn's new head coach.[40] Coach Gene Chizik then hired Gus Malzahn as the Tigers' new Offensive Coordinator.

In 2010, Auburn defeated Oregon 22–19 in the 2011 BCS National Championship Game to secure the school's second national championship. The Tigers finished the season with a 14–0 record, including comeback wins over Clemson, South Carolina, Georgia, and Alabama. The Tigers trailed the Tide 24–0 in Tuscaloosa, but managed a 28–27 comeback victory in the 75th edition of the Iron Bowl. Auburn would again defeat South Carolina 56–17 in the 2010 SEC Championship Game, claiming the school's eleventh conference championship. The Tigers were led by head coach Gene Chizik, offensive coordinator Gus Malzahn, quarterback and Heisman Trophy winner Cam Newton, and defensive tackle and Lombardi Award winner Nick Fairley.

In addition to the 1957 and 2010 championships, Auburn's 1913, 1914, 1958, 1983, 1993, and 2004 teams have also been recognized as national champions by various ranking organizations.[41]

Water tower bearing logo

## Swimming and diving

In the last decade under head coaches David Marsh, Richard Quick and co-head coach Brett Hawke, Auburn's swimming and diving program has become preeminent in the SEC and nationally, with consecutive NCAA championships for both the men and women in 2003 and 2004, then again in 2006 and 2007. Since 1982, only 8 teams have claimed national championships in women's swimming and diving. Auburn and Georgia each won nine straight(five Auburn, four Georgia) between 1999 and 2007. The men won their fifth consecutive national title in 2007, and the women also won the national title, in their case for the second straight year. The Auburn women have now won five national championships in the last six years. As of 2009, the Auburn men have won the SEC Championship fifteen out of the last sixteen years, including the last thirteen in a row, and also won eight NCAA national championships (1997, 1999, 2003, 2004, 2005, 2006, 2007 and 2009).[42] Coach Marsh, who has been a U.S. Olympic coach, is considered one of the top three swim coaches in the world, and AU swimmers have represented the U.S. and several other countries in recent Olympic Games. Auburn's most famous swimmer is Olympic gold medalist Rowdy Gaines, and also Brazilian César Cielo Filho, bronze(100m freestyle) and gold medal(50m freestyle) at the 2008 Beijing Olympic Games. As the most successful female Olympic swimmer Kirsty Coventry (swimming for her home country of Zimbabwe) who won gold, silver, and bronze medals at the 2004 Summer Olympics in Athens. While the football team is far more well-known nationally and in the media, Auburn swimming and diving is the most dominant athletics program for the university.

## Men's basketball

The Auburn men's basketball team has enjoyed off-and-on success over the years. Its best known player is Charles Barkley. Other NBA players from Auburn are John Mengelt, Rex Fredicks, Eddie Johnson, Mike Mitchell, Chuck Person, Chris Morris, Wesley Person, Chris Porter, Mamadou N'diaye, Jamison Brewer, Moochie Norris, Marquis Daniels, and Pat Burke. The Auburn University Board of Trustees approved the building of a new $92.5 million basketball arena and practice facility. Groundbreaking for the new arena occurred in the summer of 2008 with the facility opening prior to the 2010–11 season.

## Women's basketball

The Auburn University women's basketball team has been consistently competitive both nationally and within the SEC. Despite playing in the same conference as perennial powerhouse Tennessee and other competitive programs such as LSU, Georgia, and Vanderbilt, Auburn has won four regular season SEC championships and four SEC Tournament championships. AU has made sixteen appearances in the NCAA women's basketball tournament and only once, in their first appearance in 1982, have the Tigers lost in the first round. Auburn played in three consecutive National Championship games from 1988–1990 and won the Women's NIT in 2003. When Coach Joe Ciampi retired at the end of the 2003–2004 season, Auburn hired former Purdue and U.S. National and Olympic team head coach, Nell Fortner. Standout former Auburn players include: Ruthie Bolton, Vickie Orr, Carolyn Jones, Chantel Tremitiere, Monique Morehouse, and DeWanna Bonner.

## Baseball

Auburn Baseball has won six SEC championships, three SEC Tournament championships, appeared in sixteen NCAA Regionals and reached the College World Series (CWS) four times. After a disappointing 2003–2004 season, former Auburn assistant coach Tom Slater was named head coach. He was replaced in 2008 by John Pawlowski. Samford Stadium-Hitchcock Field at Plainsman Park is considered one of the finest facilities in college baseball and has a seating capacity of 4,096, not including lawn areas. In addition to Bo Jackson, Auburn has supplied several other players to Major League Baseball, including Frank Thomas, Gregg Olson, Scott Sullivan, Tim Hudson, Mark Bellhorn, Jack Baker, Terry Leach, Josh Hancock, Gabe Gross, and Steven Register.

## Women's golf

Auburn's Women's Golf team has risen to be extremely competitive in the NCAA in recent years. Since 1999, they hold a 854–167–13 (.826 win percentage) record. The team has been in five NCAA finals and finished second in 2002 and then third in 2005. The program has a total of seven SEC Championships (1989, 1996, 2000, 2003, 2005, 2006, and 2009). The seven titles is third all time for Women's golf.[43] In October 2005, Auburn was named the #3 team nationally out of 229 total teams since 1999 by *GolfWeek* magazine. Auburn's highest finish in the NCAA tournament was a tie for 2nd in 2002.[44]

Since 1996, the team has been headed by Coach Kim Evans, a 1981 alumna, who has turned the program into one of the most competitive in the nation. Coach Evans has helped develop All-Americans, SEC Players of the Year as well as three SEC Freshman of the Year. She has led the Tigers to eight-straight NCAA appearances. She is by far the winningest Coach in Auburn Golf History, having over 1100 wins and winning six of Auburn's seven total SEC Titles. Evans was named National Coach of the Year in 2003 and has coached 8 individual All-Americans while at Auburn.

## Track and field

The Auburn women's track and field team won its first ever national title in 2006 at the NCAA Outdoor Track and Field Championships, scoring 57 points to win over the University of Southern California, which finished second with 38.5 points. Auburn posted All-American performances in nine events, including two individual national champions and three second-place finishers, and broke two school records during the four-day event.

Auburn's men's team finished second at the 2003 NCAA Outdoor Championships and at the 1978, 1997 and 2003 NCAA Indoor Championships. The women's team finished 14th (2002, 2003) at the Outdoor Championships and seventh (2003) at the Indoor Championships.

## Equestrian

Auburn's Equestrian team captured the 2006 national championship, the first equestrian national championship in school history. Senior Kelly Gottfried and junior Whitney Kimble posted team-high scores in their respective divisions as the Auburn equestrian team clinched the overall national championship at the 2006 Varsity Equestrian Championships at the EXPO/New Mexico State Fairgrounds in Albuquerque, N.M. In 2008, the Auburn Equestrian team captured the 2008 Hunt Seat National Championship. Over fences riders finished 12−1−1 overall for the week. Auburn has also consistenly been highly ranked in the Women's Intercollegiate Equestrian National Coaches Poll as well.

## Fight Song

Notable among a number of songs commonly played and sung at various events such as commencement and convocation, and athletic games are: *War Eagle* the Auburn University fight song.

# Selected student organizations

## Media and publications

- *The Auburn Plainsman* – the university's student-run newspaper, has won 23 National Pacemaker Awards from the Associated Collegiate Press since 1966. Only the University of Texas' student paper has won more.[45]
- WEGL 91 FM – The Auburn campus radio station which is open to students of all majors as well as faculty and staff who wish to DJ.
- The Southern Humanities Review – One of the leading literary journals in the region, *The Southern Humanities Review* has been published at the University by members of the English faculty, graduate students in English, and the Southern Humanities Council since 1967, publishing the work of nationally known authors such as Kent Nelson and R. T. Smith.
- Eagle Eye TV News – A weekly 30-minute television news program that is produced by Auburn University students and that airs on-campus, off-campus, and on-demand at the university website.
- *The Auburn Circle* – The student general-interest magazine. *The Circle* publishes poetry, art, photography, fiction, nonfiction, and architectural and industrial design from Auburn students, faculty, staff, and alumni.
- *Glomerata* – Auburn University's student-run yearbook which began production in 1897. Its name is derived from the conglomeration of Auburn.
- Auburn University Office of Communications and Marketing – Auburn University's news outlet for media related to the accomplishments of university faculty, staff and students.
- Auburn University's official YouTube channel – Auburn University's YouTube channel was announced on January 15, 2008.[46] It contains a wide variety of videos, from promotional to educational. AU's Office of Communications and Marketing manages the content on the university's YouTube Channel.
- *www.theauburner.com* – Website written by former Auburn University Students and graduates Mark Paden and Ryan Stephens

## General interest

- Auburn University Student Space Program (AUSSP) – The AUSSP is a student-led, faculty-mentored program to design, build, launch, and operate spacecraft. Participants launch high-altitude balloons to the edge of space to test engineering and science instruments, they build small satellites that orbit Earth, and they are working with other universities on missions to the moon and Mars. The AUSSP is made of three groups: the Auburn High Altitude Balloon (AHAB) group, the AubieSat-1 [47] (Small Satellite) group, and the management group – involving students who are not majoring in the sciences or engineering. Many students take Directed Reading in Physics (PHYS 4930) and get credit for participating in AUSSP.
- United Nations World Food Programme (WFP) – Auburn University is the WFP's lead academic partner in a recently launched student "War on Hunger" campaign. In 2004, the WFP tasked Auburn University with heading the first student-led War on Hunger effort. Auburn then founded the Committee of 19 which has led campus and community hunger awareness events and developed a War on Hunger model for use on campuses across the country. The Committee of 19 recently hosted a War on Hunger Summit at which representatives from 29 universities were in attendance.
- Cooperative Education (Co-Op) – Co-op at Auburn University is a planned and supervised program alternating semesters of full-time college classroom instruction with semesters of full-time paid work assignments. These work assignments are closely related to the student's academic program. Thousands of Auburn University graduates, especially engineering majors, have supported themselves financially while studying at Auburn by participating in Co-op. This educational program prepares students for professional careers by combining academic training with practical work experience in industry, business, and government.
- The Sol of Auburn – Auburn University's Solar Car Team – recently participated in the North American Solar Challenge 2005. On July 27, 2005, Auburn's car crossed the finish line in Calgary, Alberta, Canada in 4th place in Stock Class, 12th Place overall. The SOL of Auburn is the only solar car in Alabama, and the project is organized by Auburn University's College of Engineering with a team of four faculty and over twenty undergraduate students.
- The War Eagle Flying Team (WEFT) – A student organization made up of both pilots and non-pilots. Most team members are Professional Flight Management, Aviation Management, or Aerospace Engineering majors. WEFT competes with other flying teams at the annual National Intercollegiate Flying Association (NIFA) sponsored Safety and Flight Evaluation Conference (SAFECON).
- Auburn University Computer Gaming Club – One of the oldest University Sponsored Computer Gaming Clubs in the USA. Weekly meetings and semesterly LAN parties.
- Samford Hall Clock Tower – Information on the Samford Hall Clock Tower, a well known symbol of Auburn University. Also includes information on the bell and carillon.

## See also

- Alabama Cooperative Extension System Auburn University's principal outreach organization
- Auburn University Chapel
- Donald E. Davis Arboretum
- Luther Duncan
- List of forestry universities and colleges
- The Auburn Plainsman the prize-winning student newspaper

# References

Notes

[1]  "Best Colleges: Auburn University" (http://colleges.usnews.rankingsandreviews.com/best-colleges/auburn-university-1009). US News. .

[2]  "Gogue Named 18th President of Auburn University" (http://www.ocm.auburn.edu/news_releases/18_pres.html). Communications and Marketing, Auburn University. 2007-03-22. . Retrieved 2007-09-11.

[3]  "Auburn University posts record enrollment, highest freshman ACT scores" (http://wireeagle.auburn.edu/news/3810). Wire Eagle. . Retrieved 2011-09-12.

[4]  "Quick Facts About AU" (http://www.ocm.auburn.edu/welcome/factsandfigures.html). Auburn University. 2007. . Retrieved 2007-09-10.

[5]  (http://www.ocm.auburn.edu/graphicservices/colors.html). Auburn University

[6]  http://www.auburn.edu/

[7]  "Auburn University" (http://info.alabama.gov/directory_detail.aspx?id=35). Alabama Department of Finance. . Retrieved 2007-03-24.

[8]  About Auburn (http://www.auburn.edu/admissions/auburn/) Office of Undergraduate Recruiting and University Scholarships. Retrieved on 2006-12-13

[9]  "About Auburn" (http://www.ocm.auburn.edu/welcome/aboutauburn.html) (URL). Auburn University. . Retrieved 2008-11-13.

[10]  A Short History of Auburn University, 1856–1984 (http://auburnbeta.org/campus_history/default.asp). Retrieved April 2, 2008.

[11]  Anson West, *History of Methodism in Alabama* (Nashville: Methodist Episcopal Church South, 1893), 738–739.

[12]  "The Cannon Lathe plaque" (http://www.flickr.com/photos/codelemur/5108012123/in/set-72157625100858387). . Retrieved 2011-01-17.

[13]  Auburn University's OIRA Report for 2009, Spring Semester

[14]  "America's Best Colleges" (http://www.forbes.com/top-colleges/list/). Forbes. 2011. . Retrieved October 6, 2011.

[15]  "National Universities Rankings" (http://colleges.usnews.rankingsandreviews.com/best-colleges). *America's Best Colleges 2012*. U.S. News & World Report. September 13, 2011. . Retrieved September 25, 2011.

[16]  "The Washington Monthly National University Rankings" (http://www.washingtonmonthly.com/college_guide/rankings_2011/national_university_rank.php). *The Washington Monthly*. 2011. . Retrieved August 30, 2011.

[17]  "Academic Ranking of World Universities: Global" (http://www.shanghairanking.com/ARWU2011.html). Institute of Higher Education, Shanghai Jiao Tong University. 2011. . Retrieved August 30, 2011.

[18]  "Top 400 – The Times Higher Education World University Rankings 2011–2012" (http://www.timeshighereducation.co.uk/world-university-rankings/2011-2012/top-400.html). The Times Higher Education. 2011. . Retrieved October 6, 2011.

[19]  "National Universities Rankings" (http://colleges.usnews.rankingsandreviews.com/college/national-search). *America's Best Colleges 2009*. U.S. News & World Report. 2009. . Retrieved 2009-05-18.

[20]  "Best Colleges: Best Values: National Universities" (http://colleges.usnews.rankingsandreviews.com/best-colleges/national-top-public). *America's Best Colleges 2009*. U.S. News & World Report. 2009. . Retrieved 2009-09-21.

[21]  "Best Business Schools" (http://grad-schools.usnews.rankingsandreviews.com/best-graduate-schools/top-business-schools/rankings/page+4). *America's Best Graduate Schools*. U.S. News & World Report. 2010. . Retrieved 2010-04-17.

[22]  "Best Education Programs" (http://grad-schools.usnews.rankingsandreviews.com/best-graduate-schools/top-education-schools/rankings). *America's Best Graduate Schools*. U.S. News & World Report. 2010. . Retrieved 2010-04-17.

[23]  "Best Engineering Schools" (http://grad-schools.usnews.rankingsandreviews.com/best-graduate-schools/top-engineering-schools/rankings/page+2). *America's Best Graduate Schools*. U.S. News & World Report. 2010. . Retrieved 2010-04-17.

[24]  "Best Medical Schools: Primary Care Rankings" (http://grad-schools.usnews.rankingsandreviews.com/best-graduate-schools/top-medical-schools/primary-care-rankings). *America's Best Graduate Schools*. U.S. News & World Report. 2010. . Retrieved 2010-04-17.

[25]  "Best Graduate Schools: Pharmacy" (http://grad-schools.usnews.rankingsandreviews.com/best-graduate-schools/top-pharmacy-schools/rankings). *America's Best Graduate Schools*. U.S. News & World Report. 2008. . Retrieved 2010-04-17.

[26]  "Public Affairs Rankings" (http://grad-schools.usnews.rankingsandreviews.com/best-graduate-schools/top-public-affairs-schools/public-affairs-rankings). U.S. News & World Report. 2008. . Retrieved 2010-04-17.

[27]  "Rehabilitation Counseling Ratings" (http://grad-schools.usnews.rankingsandreviews.com/best-graduate-schools/top-health-schools/rehabilitation-counseling-rankings). U.S. News & World Report. 2008. . Retrieved 2010-04-17.

[28]  "U.S. News ranks Auburn among Top 50 public universities for 15th consecutive year" (http://wireeagle.auburn.edu/news/175). Wire Eagle at Auburn University. 2007-08-17. . Retrieved 2007-08-22.

[29]  "Auburn moves up, ranks among Top 50 universities for 17th consecutive year in U.S. News survey" (http://wireeagle.auburn.edu/news/1088). Wire Eagle at Auburn University. 2009-08-20. . Retrieved 2009-09-06.

[30]  "Top Public Schools: National Universities". America's Best Colleges 2011. *U.S. News & World Report*, 2010. Retrieved: October 25, 2010. (http://colleges.usnews.rankingsandreviews.com/best-colleges/national-top-public)

[31]  Peterson, Thompson. "Auburn University Overview" (http://web.archive.org/web/20060831201554/http://www.petersons.com/ugchannel/code/instvc.asp?inunid=5220&sponsor=1&WT.mc_id=244&WT.mc_r=180&cid=vcc_newsletter#Facilities). Archived from the original (http://www.petersons.com/ugchannel/code/instvc.asp?inunid=5220&sponsor=1&WT.mc_id=244&WT.mc_r=180&cid=vcc_newsletter#Facilities) on 2006-08-31. . Retrieved 2006-09-26.

[32]  Jackson, Kristen (2006). "AU launches largest-ever fundraising campaign" (http://www.ocm.auburn.edu/news_releases/
      recordfundraise06.html). Auburn University Office of Communications and Marketing. . Retrieved 2006-08-13.
[33]  "Frequently Asked Questions" (http://mises.org/about/3467#part). Mises.org – Mises Institute. . Retrieved 2010-12-29.
[34]  Caddell, Sallie (March/April 2008). "Auburn University Aviation" (http://www.autopilotmagazine.com/articles/articleview.
      aspx?artID=1324). AutoPILOT magazine. . Retrieved 2008-06-03.
[35]  (http://www.cadc.auburn.edu/aumred/program.html)
[36]  Auburn University Campus Map – https://oitapps.auburn.edu/campusmap/click on each building for more information.
[37]  Buildings at auburn University http://www.lib.auburn.edu/architecture/buildings/
[38]  "College Search" (http://collegesearch.collegeboard.com/search/CollegeDetail.jsp?collegeId=1608&profileId=8). Auburn University –
      Housing & Campus Life. . Retrieved 2010-12-29.
[39]  "Board approves tuition increase, new residence halls" (http://www.theplainsman.com/front/
      board_approves_tuition_increase_new_residence_halls). *The Auburn Plainsman*. . Retrieved 2007-04-29.
[40]  "Auburn hires Gene Chizik as football coach" (http://www.sportingnews.com/yourturn/viewtopic.php?t=496642). Sporting News.
      2008-12-13. . Retrieved 2008-12-13.
[41]  College Football Data Warehouse, *Yearly National Championship Selectors*, 1913 (http://www.cfbdatawarehouse.com/data/
      national_championships/yearly_results.php?year=1913), 1914 (http://www.cfbdatawarehouse.com/data/national_championships/
      yearly_results.php?year=1914), 1958 (http://www.cfbdatawarehouse.com/data/national_championships/yearly_results.php?year=1958),
      1983 (http://www.cfbdatawarehouse.com/data/national_championships/yearly_results.php?year=1983), 1993 (http://www.
      cfbdatawarehouse.com/data/national_championships/yearly_results.php?year=1993), 2004 (http://www.cfbdatawarehouse.com/data/
      national_championships/yearly_results.php?year=2004). Retrieved July 28, 2008.
[42]  "Auburn Men's Swimming and Diving Makes it Four-in-a-Row, Complete Sweep of 2006 NCAA Swimming and Diving Titles" (http://
      auburntigers.collegesports.com/sports/c-swim/recaps/032506aaa.html). Auburn University and CSTV Networks, Inc.. 2006-03-15. .
      Retrieved 2007-09-10.
[43]  "No. 6 Women's Golf Rallies In Final Round To Win SEC Championship" (http://auburntigers.cstv.com/sports/w-golf/recaps/
      041909aaa.html). Auburn University. . Retrieved 2009-05-30.
[44]  "Auburn Women's Golf named No. 3 Program since 1999" (http://auburntigers.cstv.com/sports/w-golf/spec-rel/101005aaa.html).
      Auburn University. . Retrieved 2009-05-30.
[45]  "Auburn university news" (http://www.auburn.edu/administration/univrel/news/archive/11_02news/11_02pacemaker.html). Auburn
      university news. 2002-11-08. . Retrieved 2008-11-18.
[46]  "AU launches Youtube channel" (http://wireeagle.auburn.edu/news/261). Auburn university news. 2008-01-15. . Retrieved 2008-11-18.
[47]  "AubieSat-1" (http://space.auburn.edu/aubiesat-1/). Space.auburn.edu. . Retrieved 2010-12-29.

## External links

- Auburn University official site (http://www.auburn.edu/)
- Auburn Tigers Athletics official site (http://www.auburntigers.com/)
- Auburn University article in the Encyclopedia of Alabama (http://www.encyclopediaofalabama.org/face/
  Article.jsp?id=h-1649)

# Auburn_Tigers_football

<table>
<tr><td colspan="2" align="center">Auburn Tigers</td></tr>
<tr><td colspan="2" align="center">Current season</td></tr>
<tr><td colspan="2"></td></tr>
<tr><td>First season</td><td>1892</td></tr>
<tr><td>Athletic director</td><td>Jay Jacobs</td></tr>
<tr><td>Head coach</td><td>Gene Chizik</td></tr>
<tr><td></td><td>3rd year, 30–10–0  ()</td></tr>
<tr><td>Other staff</td><td>Trooper Taylor (assistant HC)<br>TBA (OC)<br>Brian VanGorder (DC)</td></tr>
<tr><td>Home stadium</td><td>Jordan-Hare Stadium</td></tr>
<tr><td>Field</td><td>Pat Dye Field</td></tr>
<tr><td>Year built</td><td>1939</td></tr>
<tr><td>Stadium capacity</td><td>87,451</td></tr>
<tr><td>Stadium surface</td><td>Grass</td></tr>
<tr><td>Location</td><td>Auburn, Alabama</td></tr>
<tr><td>League</td><td>NCAA Division I (FBS)</td></tr>
<tr><td>Conference</td><td>SEC</td></tr>
<tr><td>Division</td><td>West (1992–present)</td></tr>
<tr><td>Past conferences</td><td>Independent (1892-1894)<br>SIAA (1895-1920)<br>Southern (1921-1932)</td></tr>
<tr><td>All-time record</td><td>712–400–47 ()</td></tr>
<tr><td>Postseason bowl record</td><td>22–13–2</td></tr>
<tr><td>Claimed national titles</td><td>2 (1957, 2010)</td></tr>
<tr><td>Conference titles</td><td>11 / 7 SEC (1900, 1913, 1919, 1932, 1957, 1983, 1987, 1988, 1989, 2004, 2010)</td></tr>
<tr><td>Division titles</td><td>7 (1997, 2000, 2001, 2002, 2004, 2005, 2010)</td></tr>
<tr><td>Heisman winners</td><td>3</td></tr>
<tr><td>Consensus All-Americans</td><td>66[1]</td></tr>
<tr><td colspan="2">Current uniform</td></tr>
</table>

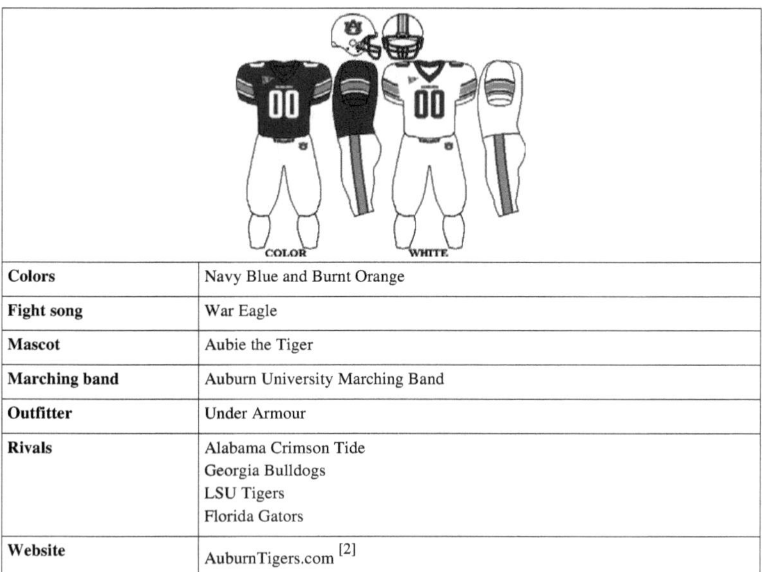


| Colors | Navy Blue and Burnt Orange |
| --- | --- |
| Fight song | War Eagle |
| Mascot | Aubie the Tiger |
| Marching band | Auburn University Marching Band |
| Outfitter | Under Armour |
| Rivals | Alabama Crimson Tide<br>Georgia Bulldogs<br>LSU Tigers<br>Florida Gators |
| Website | AuburnTigers.com [2] |

The **Auburn Tigers football** team represents Auburn University in the Football Bowl Subdivision of the National Collegiate Athletics Association (NCAA) and the Western Division of the Southeastern Conference. Since its beginnings in 1892, Auburn football has accumulated 2 national championships, 11 conference championships, 7 divisional championships, 7 perfect seasons, and 3 Heisman Trophy winners. Auburn is the thirteenth winningest football program in FBS history, and has compiled over 700 victories, while appearing in 37 postseason bowl games and ranking fifth nationally for bowl winning percentage. Auburn's home stadium is Jordan-Hare Stadium, which becomes Alabama's fifth largest city on gamedays with a capacity of 87,451. Auburn's archrival is in-state foe Alabama. The Tigers and Crimson Tide meet annualy in the Iron Bowl, one of the biggest rivalries in all of sports. The Tigers also maintain rivalries with SEC foes Georgia and LSU. Auburn's winningest coach is Ralph "Shug" Jordan, who led the Tigers from 1951 to 1975, and won the school's first national championship in 1957. The team is currently coached by Gene Chizik, who led Auburn to its second national championship in 2010.

## Beginnings (1892-1903)

The organization of Auburn's first football team is credited to George Petrie, who led the 1892 Tigers to a 2-2 record. Petrie also chose burnt orange and navy blue as the official colors for Auburn athletic teams, which was inspired by his alma mater, the University of Virginia. The first game was against the University of Georgia at Piedmont Park in Atlanta, Georgia. Auburn won 10-0 in front of a crowd of 2,000, in a game that would establish the Deep South's Oldest Rivalry. Auburn met in-state rival Alabama for the first time ever at Lakeview Park in Birmingham, Alabama during the 1893 season, which ended with a 32-22 victory for the Tigers. Auburn was led by nine different coaches over a 12-year span, which most notably included John Heisman (1895–1899), for whom the Heisman trophy is named. During five years, Heisman compiled a 12-4-2 record, before departing for Clemson in 1900. Auburn's first conference membership came in 1895, when it joined the Southern Intercollegiate Athletic Association (SIAA). The first conference championship and undefeated season came in 1900, when the Tigers went 4-0 under coach Billy Watkins.

## Mike Donahue era (1904-1922)

In 1904, Mike Donahue became the tenth head coach of the Auburn Tigers. His first team finished with a 5-0 record, marking Auburn's second undefeated season. The 1907 season would mark the last time Auburn would play Alabama until 1948, after a 7-7 tie between the two rivals. Donahue's most notable season came in 1913, when Auburn claimed its second conference championship with an 8-0 record. The Tigers were retroactively awarded a national title by numerous polls, including Billingsley Report, James Howell, and 1st-N-Goal. The 1914 team also won a conference championship with an 8-0-1 record, and was retroactively awarded a national title by James Howell. The Tigers would return to the top of the conference once again in 1919, with an 8-1 record. Auburn moved to the Southern Conference in 1921, one year before Donahue's departure from Auburn in 1922, before becoming the coach at LSU in 1923.

## Searching for success (1923-1950)

Auburn experienced a decline in success during the late 1920s, 1930s, and 1940s. The Tigers underwent nine different coaching changes over a 28-year period. The only conference championship during this time came in 1932, after a 9-0-1 season under coach Chet A. Wynne. In 1933, Auburn, along with 12 other institutions, left the Southern Conference to form the Southeastern Conference (SEC). Auburn's first bowl appearance came in 1936 under coach Jack Meagher after a 7-2-2 season. The Tigers traveled to Havana, Cuba to play Villanova in the Bacardi Bowl, which ended in a 7-7 tie. Auburn's first bowl win came after the 1937 season against Michigan State in the Orange Bowl. Due to the events surrounding World War II, Auburn did not field a team in 1943, but resumed competition in 1944 under Carl Voyles. During Earl Brown's tenure, Auburn met Alabama for the first time since 1907, which ended with an Alabama victory. The Tigers quickly responded in 1949, as they stunned the heavily-favored Crimson Tide in a 14-13 victory. An 0-10 season in 1950 called for a change, and marked the end of a trying era for Auburn football.

## Ralph "Shug" Jordan era (1951-1975)

In 1951, Auburn hired Ralph "Shug" Jordan to become the new head coach of the Tigers. During his first season, Auburn finished with a 5-5 record. He led the Tigers to three consecutive bowl appearances in 1953, 1954, and 1955. Jordan is most recognized for his 1957 squad, which finished the season with a 10-0 record, and won Auburn's first SEC Championship. The Associated Press named the Auburn Tigers no. 1 in its postseason poll, which marked the first official national championship in school history. Auburn was ineligible to participate in postseason play during 1957 and 1958 seasons due to NCAA sanctions for illegal recruiting inducements. The 1958 team was also named national champions by Montgomery Full Season Championship poll, after a 9-0-1 season. Auburn went on to appear in bowl games in 1963 and 1965. Beginning in 1968, the Tigers enjoyed seven consecutive bowl appearances under coach Jordan. In 1971, Auburn quarterback Pat Sullivan led the Tigers to a 9-2 record, and became the school's first Heisman Trophy winner. Auburn would go on to lose the 1972 Sugar Bowl to Oklahoma, 40-22. One of Jordan's biggest victories came against Alabama in 1972, when the Tigers shocked the Crimson Tide in a 17-16 upset. The 1972 Iron Bowl became known as the "Punt Bama Punt" game, due to two blocked Alabama punts in the fourth quarter, which were both returned for Auburn touchdowns. In 1973, Auburn's Cliff-Hare Stadium was renamed Jordan-Hare Stadium, which was the first stadium in the nation to be named for an active coach. After the 1975 season, Jordan retired after a 25-year tenure at Auburn. He has remained Auburn's winningest coach, with a 176-83-7 record, and a .675 winning percentage.

## Struggle for respect (1976-1980)

Following Jordan's retirement, Auburn hired Doug Barfield to become the new head coach. From 1976 to 1980, Barfield's Tigers compiled a 27-27-1 on-field record, with no bowl appearances. He lost all five games to rival Alabama during his tenure, and was later awarded two victories due to forfeits by Mississippi State in 1976 and 1977, making his record 29-25-1. He was dismissed from his position after a disappointing season in 1980, as the Tigers finished with a 5-6 record. Auburn then hired Pat Dye, a former assistant coach at Alabama under Coach Paul W. Bryant, and head coach at Wyoming at the time. During his first season in 1981, Auburn finished with a 5-6 record.

## Pat Dye era (1981-1992)

In 1982, Pat Dye led Auburn to a 9-3 record and its first bowl appearance in eight years. The 1982 season would also begin a streak of nine consecutive bowl game appearances. The most notable game of the season came against Alabama in the Iron Bowl, when Auburn snapped the Tide's 9-game winning streak. The 1982 Iron Bowl is widely known as the "Bo Over the Top" game, for Auburn running back Bo Jackson's leap over the top of a pile from the one yard line to secure a 23-22 victory over Alabama. This would be the final Iron Bowl for Alabama's legendary coach, Paul W. Bryant, who retired after the 1982 season. Dye's most storied season came in 1983, when the Tigers went 11-1, claiming the conference championship. Auburn went on to defeat Michigan in the Sugar Bowl 9-7. Some felt that #3 Auburn should have been crowned the national champions, due to #5 Miami's upset of #1 Nebraska in the Orange Bowl, and #7 Georgia's upset of #2 Texas in the Cotton Bowl. Nonetheless, Miami jumped from No. 5 to No. 1 in both the AP and Coaches polls, while Auburn remained in the No. 3 spot behind #2 Nebraska. The Tigers were named national champions by various polling organizations, such as the New York Times and Billingsley Report. In 1985, running back Bo Jackson would become the school's second Heisman Trophy winner. Auburn would go on to win three consecutive SEC Championships in 1987, 1988, and 1989. In 1988, defensive tackle Tracy Rocker became the school's first Lombardi Award winner and also won the Outland Trophy. Pat Dye is credited for organizing the first ever Iron Bowl played in Auburn. On December 2, 1989, Bill Curry's #2 Crimson Tide (10-0) traveled to Jordan-Hare Stadium, which had surpassed the seating capacity of Legion Field, to face the #11 Auburn Tigers, who defeated the Tide 30-20. The 1989 Iron Bowl would continue a 4-game winning streak over Alabama. Since 1981, Auburn has a 17-13 edge over Alabama in Iron Bowl wins. Dye's tenure on the plains ended when Auburn was penalized for payments by boosters and assistant coaches to a player, Eric Ramsey. Tape recordings were released that implicated a booster named "Corky" Frost, and present Troy University head coach Larry Blakeney. The controversy landed the Auburn program a spot on *60 Minutes* and an eventual NCAA investigation. While the investigation did not find Dye personally responsible for rules violations, the NCAA determined that as head coach and athletic director, Dye should have known about and stopped the payments to Ramsey. The fallout from the NCAA probation against the football team pushed Dye out as athletic director in 1991 and as head coach the following year. Over twelve seasons, Dye achieved a 99-39-4 record, making him the third winningest coach in Auburn football history, only behind Mike Donahue and Ralph "Shug" Jordan. In 2005, the playing surface of Jordan-Hare Stadium was named "Pat Dye Field" in honor of Dye's achievements and contributions he made to Auburn during his tenure.

## Terry Bowden era (1993-1998)

Following the departure of Pat Dye, Auburn named Terry Bowden, son of legendary coach Bobby Bowden, head coach of the Tigers. In 1993, while serving a one-year television ban and two-year postseason bowl ban due to NCAA probation, Auburn shocked the nation by completing the season with a perfect 11-0 record. The Tigers were not eligible to play in the SEC Championship Game, nor a bowl game, but were named national champions by the National Championship Foundation. The most memorable game, of the 1994 season was the "Interception Game" versus LSU. In which the Auburn defence intercepted 7 LSU passes, returning 3 for touchdowns in the 4th quarter

(Ken Alvis, Fred Smith and Brian Robinson). During the first 2 seasons under Bowden, the Tigers amassed a 20-1-1 record. After serving two years of probation, Auburn made three consecutive bowl game appearances from 1995 to 1997. Bowden's 1997 team won the SEC Western Division title, and played in the SEC Championship Game, falling to Tennessee 30-29. In 1998, Bowden faced criticism for recruiting woes, off-the-field issues, and player discipline, which eventually led to his resignation after a 1-5 start on the season. Interim head coach Bill Oliver finished out the season, which ended with a 3-8 record. Bowden compiled a 47-17-1 record at Auburn after six seasons as head coach.

## Tommy Tuberville era (1999-2008)

Following the 1998 season, Ole Miss head coach Tommy Tuberville left Oxford to become the new head coach of the Auburn Tigers. In his first season, the Tigers finished with a 5-6 record, but would return to the SEC Championship Game in 2000, following a 9-0 victory over Alabama, which was played in Tuscaloosa for the first time in 99 years. The Tigers fell to Florida 28-6, but would begin a streak of eight consecutive bowl appearances. Auburn would win a share of the SEC Western Division title in 2001 and 2002. His 2002 season is most notably known for Auburn's 17-7 upset victory over Alabama, which began a six-year winning streak over the Tide. Tuberville's 2004 team completed the season with a perfect 13-0 record and an SEC Championship. Auburn was left out of the BCS National Championship Game, due to its low preseason ranking and the presence of two other undefeated teams ranked higher, #1 USC (12-0) and #2 Oklahoma (12-0). The Tigers went on to defeat Virginia Tech 16-13 in the Sugar Bowl, completing Auburn's third perfect season in the modern era of college football. USC defeated Oklahoma 55-19 to win the national championship, while Auburn was ranked No. 2 in the final AP and Coaches polls. The Tigers were recognized as national champions by various polling organizations, including FansPoll and Golf Digest. Tuberville came under much criticism during the 2008 season for his lackluster performance and coaching staff, including offensive coordinator Tony Franklin, whom he fired after a shocking 14-13 loss to Vanderbilt in October. Auburn finished the year with a 5-7 record, after a disappointing 36-0 loss to rival Alabama in the Iron Bowl, marking the Tide's first victory over Auburn in Tuscaloosa and snapping Auburn's six-year winning streak. Tuberville voluntarily resigned the following week, stating that he would take a year off from coaching. Over ten seasons, Tuberville compiled an 85-40 record at Auburn, while winning one conference championship, five division championships, and completing Auburn's third perfect season in modern history.

## Gene Chizik era (2009-present)

On December 13, 2008, Athletic Director Jay Jacobs announced Gene Chizik, former Auburn defensive coordinator and then Iowa State head coach, as the new Auburn head coach. He received early criticism for his 5-19 record during his time at Iowa State during 2007 and 2008. He quickly began forming his new coaching staff, including offensive coordinator Gus Malzahn, who had coached the nation's top offense at Tulsa for the past two seasons. During his first season, Auburn finished with a 7-5 record, and defeated Northwestern 38-35 in the Outback Bowl, its first bowl game since 2007. Following the 2009 season, Chizik and his staff secured a top-5 recruiting class, highlighted by junior college transfer quarterback Cam Newton and running back Mike Dyer. Auburn's 2010 "A-Day" spring scrimmage drew a crowd of 63,217 fans to Jordan-Hare Stadium, setting a new spring game attendance record. Auburn, led by quarterback Cam Newton, running back Mike Dyer, and defensive tackle Nick Fairley, completed the regular season with a perfect 12-0 record, highlighted by a comeback victory over Alabama. The Tide led Auburn 24-0 in the first half, only to lose the game in the second half 28-27. It was the largest lead ever blown by Alabama in Tuscaloosa. Auburn went on to defeat South Carolina 56-17 in the SEC Championship Game, which secured a spot in the BCS National Championship Game. This would be the first BCS bowl game appearance for Auburn since 2004, when the Tigers were left out of the national championship picture. Cam Newton became the third Heisman Trophy winner in school history, while also winning the AP Player of the Year Award, the Walter Camp Award, the Davey O'Brien Award, the Manning Award, and the Maxwell Award. Nick Fairley became the second Auburn player in school history to win the Lombardi Award. Auburn faced the Oregon Ducks on January 10, 2011 in Glendale, Arizona, which ended with a 22-19 Auburn victory, secured by a game-winning field goal kick by senior Wes Byrum, who also kicked the game-winning field goals against Clemson and Kentucky during the regular season. Auburn finished the season with a perfect 14-0 record, and its first official national championship since 1957. Auburn celebrated their national championship with a special ceremony at Jordan-Hare Stadium two weeks following the championship game in Arizona. The coaches and players were honored, along with players from the 1957, 1993, and 2004 undefeated teams. The event drew over 78,000 fans, covering Pat Dye Field and the lower bowl, spilling into both upper decks. A special "reverse" Tiger Walk and special rolling of Toomer's Corner also took place. After settling down from the magical 2010 season, Chizik and his staff began preparing to defend their national title. Auburn opened the 2011 season with dramatic wins against Utah State and Mississippi State. Auburn then fell to eventual ACC Champion Clemson on the road in Death Valley, which snapped Auburn's 17-game winning streak (began on January 1, 2010, vs. Northwestern in Outback Bowl). The Tigers would go on to complete the regular season with a 7-5 record and ranked no. 25 in the final BCS poll, with wins against Florida Atlantic, no. 9 South Carolina, Florida, Ole Miss, and Samford. Auburn fell to Arkansas, LSU (SEC Champions), Georgia (eastern division champion), and archrival Alabama (National Champions). The Tigers won their 37th bowl appearance by a score of 43-24 over the Virginia Cavaliers in the 2011 Chick-fil-A Bowl on December 31, 2011.

## Historical ranking

Auburn has the 13th most wins in D-1A college football.[3] In terms of winning percentage, Auburn ranks as the 9th most successful team in the past 25 years with a 71% win rate (213-86-5)[4] and 9th over the last half century (1955–2010) with 69%.[5] Of the 93 current I-A football programs that been active since Auburn first fielded a team 116 years ago, Auburn ranks 15th in winning percentage over that period.[6]

The College Football Research Center lists Auburn as the 14th best college football program in history,[7] with eight Auburn squads listed in Billingsley's Top 200 Teams of All Time (1869–2010).[8] After the 2008 season, ESPN ranked Auburn the 21st most prestigious program in history.[9]

The Associated Press poll statistics show Auburn with the 11th best national record of being ranked in the final AP Poll[10] and 13th overall (ranked 496 times out of 1021 polls since the poll began in 1936), with an average ranking of 11.02.[11] Since the Coaches Poll first released a final poll in 1950, Auburn has 26 seasons where the team finished ranked in the top 20 in both the AP and Coaches Polls.[12]

## Heisman links

Three Auburn players, Pat Sullivan in 1971, Bo Jackson in 1985, and Cam Newton in 2010, have won the Heisman Trophy. The Trophy's namesake, John Heisman, coached at Auburn from 1895 until 1899. Of the eight schools of which Heisman coached (among others, Georgia Tech and Clemson), Auburn is the only school that has produced a Heisman Trophy winner. The Auburn Athletic Department has announced that it will honor the school's three Heisman winners with statues, along with a bust of coach John Heisman, outside the east side of Jordan-Hare Stadium.[13]

## Modern history

While Auburn football has a long and storied history, the Tigers have been quite successful in recent years. Since the expansion of the SEC and the split into divisions, Auburn has been the winningest SEC West team in league play since the conference realignment in 1992.[14] As of the end of the 2010 season, Auburn teams have won 41 of their last 58 conference matchups including 20 of the last 28 SEC away games. The Tigers seem to perform best when facing their greatest challenge as, in addition to the success on the road in the SEC, Auburn teams have won 12 of their last 19 matchups versus top-10 opponents. The Tigers also have done well protecting Jordan-Hare Stadium, particularly at night where the home team has won 21 of the 25 night games since 2000. Over the five previous seasons, Auburn won 47 games (72.3%).

## 1983 season

The 1983 Auburn Tigers, led by head coach Pat Dye and running back Bo Jackson, finished 11-1 after playing the nation's toughest schedule. Their only loss came against #3 Texas, who defeated the Tigers 20-7. Auburn went on to defeat #8 Michigan 9-7 in the Sugar Bowl. Despite entering the bowl games ranked third in both major polls, and with both teams ranked higher losing their bowl games, the Tigers ended ranked third in the final AP poll. This despite the fact that Miami had played a dead average schedule while Auburn had played one of the toughest schedules in Championship contender history. Auburn received 21 polls to Miami's 14.

## 1993 season

Head coach Terry Bowden led the 1993 team to a perfect season in his first year on the Plains. The Tigers were the only undefeated team in major college football, however were banned from playing on television or post-season games due to NCAA violations. Rival Alabama was sent to the SEC Championship Game as the substitute representative of the Western Division. Auburn finished ranked fourth in the nation by the Associated Press but was named a co-National Champion by the National Championship Foundation along with Florida State, Nebraska and Notre Dame.

## 2004 season

The Auburn Tigers ended the 2004 season undefeated, but were left out of the BCS title game because they ranked third in the final BCS rankings. That left undefeated USC and Oklahoma (ranked No. 1 and No. 2 respectively) to play in the Orange Bowl for the National Championship. Auburn went on to win the Sugar Bowl against Virginia Tech. The team finished No. 2 in both the final AP Poll and USA Today Coaches Poll, following Oklahoma's loss in the National Championship game to the University of Southern California (USC). USC was later stripped of the National Championship for violating NCAA rules.

## 2010 season

On October 24, 2010, Auburn was ranked first in the BCS polls for the first time in school history. Their quarterback, Cameron Newton, became a Heisman Trophy winner. He had a total of 2,854 yards passing and 30 passing touchdowns. He also rushed for 1,473 yards and 20 touchdowns. Auburn ended the 2010 regular season a perfect 12-0 after a come-back win over cross-state rival, Alabama, in the 75th "Iron Bowl" game by a score of 28-27. Auburn defeated South Carolina in the December 4th, 2010 SEC Championship game in Atlanta, GA by a score of 56 -17. This SEC Championship victory clinched Auburn's berth in the BCS National Championship Game which took place in Glendale, Arizona against the Oregon Ducks. The Tigers defeated the Ducks 22-19 with a last-second field goal to win their second national championship.

# Team Awards and records

## National championships

Nine Auburn teams have been awarded some form of "National Champions" title, though Auburn only officially claims two—a share of the 1957 title, awarded by the Associated Press;[15] and an undisputed national championship in 2010.

| Year | Coach | Selector | Record | Bowl | Result |
|---|---|---|---|---|---|
| 1957 | Ralph "Shug" Jordan | AP | 10-0 (7-0) | *no bowl* | **Auburn 40**,Alabama 0 |
| 2010 | Gene Chizik | BCS, AP, Coaches | 14-0 (8-0) | BCS National Championship Game | **#1 Auburn 22**, #2 Oregon 19 |
| Claimed national championships: | | | | 2 | |

Over the seasons, other polls and organizations have named Auburn national champions,[16] although none of these championships are recognized by the school:

* 1913, 1914, 1958, 1983, 1993, 2004

The AP Poll did not begin selecting a champion until 1936 nor the AFCA Coaches Poll until 1950, so many national champion titles previous to those date were awarded retroactively. The undefeated 1913 and 1914 teams coached by Mike Donahue were some of the best defenses in Auburn history. In fact, the 1914 squad allowed zero points all season, outscoring opponents 193-0. The 1983 team featuring Bo Jackson went 11−1 and finished the season by beating Michigan in the Sugar Bowl. Despite the team entering the game ranked third in the AP and both teams ranked ahead losing their bowl games, Auburn was jumped by fifth ranked Miami for the AP National Title. The 1993 team was ineligible to play in a postseason bowl game due to NCAA-imposed sanctions for paying Eric Ramsey[17] and finished ranked fourth by the AP. The undefeated 2004 squad (13−0) finished second in the AP and Coaches Top 25 polls, but the team was awarded the 2004 Fanspoll.com People's National Champion title.[18] After USC was stripped of the FWAA title, the organization discussed awarding the Grantland Rice Award to Auburn but ultimately voted not to award a trophy for 2004.[19] On June 6, 2011, the BCS officially stripped USC of its 2004 national championship, stating that there would be no champion for the 2004 season. Nonetheless, Auburn acknowledges the 2004 title, along with the 1913, 1983 and 1993 titles, in its media guide.[20]

## Undefeated seasons

Since its beginnings in 1892, Auburn has completed twelve undefeated seasons.[21] This includes seven perfect seasons in which the Tigers were undefeated and untied:

* 1900, 1904, 1913, 1957, 1993, 2004, 2010

## Conference championships

Auburn has won a total of 11 conference championships, including 7 SEC Championships.

**Conference affiliations:**

* 1892-1894, Independent
* 1895-1920, Southern Intercollegiate Athletic Association
* 1921-1932, Southern Conference
* 1933–present, Southeastern Conference

| Year | Coach | Conference | Record | Bowl | Result |
|------|-------|-----------|--------|------|--------|
| 1900 | Billy Watkins | SIAA | 4-0 | *no bowl* | |
| 1913 | Mike Donahue | SIAA | 9-0 | *no bowl* | |
| 1919 | Mike Donahue | SIAA | 8-0-1 | *no bowl* | |
| 1932 | Chet A. Wynne | Southern | 9-0-1 | *no bowl* | |
| 1957 | Ralph "Shug" Jordan | SEC | 10-0 (7-0) | *no bowl* | |
| 1983 | Pat Dye | SEC | 11-1 (6-0) | Sugar Bowl | **#3 Auburn 9**, #8 Michigan 7 |
| 1987 | Pat Dye | SEC | 9-1-2 (6-0-1) | Sugar Bowl | *#6 Auburn 16, #4 Syracuse 16* |
| 1988 | Pat Dye | SEC | 10-2 (6-1) | Sugar Bowl | #7 Auburn 7, **#4 Florida State 13** |
| 1989 | Pat Dye | SEC | 10-2 (6-1) | Hall of Fame Bowl | **#9 Auburn 31**, #20 Ohio State 14 |
| 2004 | Tommy Tuberville | SEC | 13-0 (8-0) | Sugar Bowl | **#3 Auburn 16**, #9 Virginia Tech 13 |
| 2010 | Gene Chizik | SEC | 14-0 (8-0) | BCS National Championship Game | **#1 Auburn 22**, #2 Oregon 19 |
| **Total Conference Championships:** | | | | **11** | |

## Divisional championships

Since 1992, Auburn has won the SEC Western Division championship outright on four occasions, and is 2-2 in the SEC Championship Game. The most recent appearance came in 2010, as Auburn completed the regular season 12-0, and defeated South Carolina 56-17 in the 2010 SEC Championship Game. Auburn has also shared the western division title, but did not play in the championship game, on three occasions.

| Year | Coach | Record | Championship Game Result |
|------|-------|--------|--------------------------|
| 1997 | Terry Bowden | 10-2 (6-2) | #11 Auburn 29, **#3 Tennessee 30** |
| 2000 | Tommy Tuberville | 9-4 (6-2) | #18 Auburn 6, **#7 Florida 28** |
| 2001 | Tommy Tuberville | 7-5 (5-3) | *co-champions* |
| 2002 | Tommy Tuberville | 9-4 (5-3) | *co-champions* |
| 2004 | Tommy Tuberville | 13-0 (8-0) | **#3 Auburn 38**, #15 Tennessee 28 |
| 2005 | Tommy Tuberville | 9-3 (7-1) | *co-champions* |
| 2010 | Gene Chizik | 14-0 (8-0) | **#1 Auburn 56**, #19 South Carolina 17 |
| **Totals** | | **7** | **2-2** |

## Rivalries

### Primary Auburn Football Rivalries: All-Time Records

| Name of Rivalry | Rival | Games Played | First Meeting | Last Meeting | AU Won | AU Lost | Ties | Streak | Latest win |
|---|---|---|---|---|---|---|---|---|---|
| Iron Bowl | Alabama | 76 | 1893 | 2011 | 34 | 41 | 1 | 1 loss | 2010, 28-27 |
| Deep South's Oldest Rivalry | Georgia | 115 | 1892 | 2011 | 54 | 53 | 8 | 1 loss | 2010, 49-31 |
| Tiger Bowl | LSU | 46 | 1901 | 2011 | 20 | 25 | 1 | 1 loss | 2010, 24-17 |
| Auburn–Florida football rivalry | Florida | 83 | 1912 | 2011 | 43 | 38 | 2 | 3 wins | 2011, 17-6 |
| Totals | | 320 | | | 151 | 157 | 12 | | |

## Traditions

### Tiger Walk

Before each Auburn home football game, thousands of Auburn fans line Donahue Drive to cheer on the team as they walk from the Auburn Athletic Complex to Jordan-Hare Stadium. The tradition began in the 1950s when groups of kids would walk up the street to greet the team and get autographs. During the tenure of coach Doug Barfield, the coach urged fans to come out and support the team, and thousands did. Today the team walks down the hill and into the stadium surrounded by fans who pat them on the back and shake their hands as they walk. The largest Tiger Walk occurred on December 2, 1989, before the first ever home football game against rival Alabama—the Iron Bowl. On that day, an estimated 20,000 fans packed the one block section of road leading to the stadium. According to former athletic director David Housel, Tiger Walk has become "the most copied tradition in all of college football."[22]

### "War Eagle"

There are many stories surrounding the origins of Auburn's battle cry, "War Eagle." The most popular account involves the first Auburn football game in 1892 between Auburn and the University of Georgia. According to the story, in the stands that day was an old Civil War soldier with an eagle that he had found injured on a battlefield and kept as a pet. The eagle broke free and began to soar over the field, and Auburn began to march toward the Georgia end-zone. The crowd began to chant, "War Eagle" as the eagle soared. After Auburn won the game, the eagle crashed to the field and died but, according to the legend, his spirit lives on every time an Auburn man or woman yells "War Eagle!" The battle cry of "War Eagle" also functions as a greeting for those associated with the University. For many years, a live golden eagle has embodied the spirit of this tradition. The eagle

Nova, "War Eagle VII"

was once housed on campus in The A. Elwyn Hamer Jr. Aviary (which was the second largest single-bird enclosure in the country), but the aviary was taken down in 2003 and the eagle moved to a nearby raptor center. The eagle, War Eagle VI (nicknamed "Tiger"), was trained in 2000 to fly free around the stadium before every home game to the delight of fans. The present eagle, War Eagle VII (nicknamed "Nova"), continues the tradition.

## Toomer's Corner

The intersection of Magnolia and College streets in Auburn, which marks the transition from downtown Auburn to the university campus, is known as Toomer's Corner. It is named after Toomer's Drugs, a small store on the corner that has been an Auburn landmark for over 150 years. Hanging over the corner are two massive old-growth oak trees, and anytime anything good happens concerning Auburn, toilet paper can usually be found hanging from the trees. Also known as "rolling the corner," this tradition is thought to have originated in the 1970s and until the mid 1990s was relegated to only to celebrating athletic wins.

## Wreck Tech Pajama Parade

The Wreck Tech Pajama Parade originated in 1896, when a group of mischievous Auburn ROTC cadets, determined to show up the more well-known engineers from Georgia Tech, snuck out of their dorms the night before the football game between Auburn and Tech and greased the railroad tracks. According to the story, the train carrying the Georgia Tech team slid through town and didn't stop until it was halfway to the neighboring town of Loachapoka, Alabama. The Georgia Tech team was forced to walk the five miles back to Auburn and, not surprisingly, were rather weary at the end of their journey. This likely contributed to their 45–0 loss. While the railroad long ago ceased to be the way teams traveled to Auburn and students never greased the tracks again, the tradition continues in the form of a parade through downtown Auburn. Students parade through the streets in their pajamas and organizations build floats.[23]

# Current Coaching Staff

| Name | Position |
| --- | --- |
| Gene Chizik | Head Coach |
| Trooper Taylor | Assistant Head Coach/Wide Receivers Coach |
| Vacant | Offensive Coordinator/Quarterbacks Coach |
| Brian VanGorder | Defensive Coordinator/Linebackers Coach |
| Jay Boulware | Special Teams Coordinator/Tight Ends Coach |
| Curtis Luper | Recruiting Coordinator/Running Backs Coach |
| Jeff Grimes | Offensive Line Coach |
| Mike Pelton | Defensive Line Coach |
| Tommy Thigpen | Safeties Coach |
| Phillip Lolley | Cornerbacks Coach |

# Head coaches

Auburn has had 25 head coaches, and 1 interim head coach, since it began play during the 1892 season.[24] Since December 2008, Gene Chizik has served as Auburn's head coach.[25] The team has played more than 1,150 games over 119 seasons.[24] In that time, six coaches have led the Tigers in postseason bowl games: Jack Meagher, Ralph Jordan, Pat Dye, Terry Bowden, Tommy Tuberville and Gene Chizik.[26] Billy Watkins, Mike Donahue, Chet A. Wynne, Jordan, Dye, Tuberville and Chizik won a combined eleven conference championships.[27] During their tenures, Jordan and Chizik each won national championships with the Tigers.[27] [28]

# Award winners

A number of Auburn players and coaches have won national awards, including 66 players being named as college football All-Americans. The Tigers also have eleven coaches and players that have been inducted into the College Football Hall of Fame in South Bend, Indiana.

## Hall of Fame

| Players<br>*Year Inducted* | Coaches<br>*Year Inducted* |
| --- | --- |
| **1954** - Jimmy Hitchcock | **1951** - "Iron Mike" Donahue |
| **1956** - Walter Gilbert | **1954** - John Heisman |
| **1991** - Pat Sullivan | **1982** - Ralph "Shug" Jordan |
| **1994** - Tucker Frederickson | **2005** - Pat Dye |
| **1998** - Bo Jackson | |
| **2002** - Terry Beasley | |
| **2004** - Tracy Rocker | |
| **2009** - Ed Dyas | |

## National Awards

### Players

| Heisman Trophy[29]<br>*Best player* | Walter Camp Award[30]<br>*Best player* | Maxwell Award<br>*Best player* |
| --- | --- | --- |
| **1971** - Pat Sullivan, *QB* | **1971** - Pat Sullivan, *QB* | **2010** - Cam Newton, *QB* |
| **1985** - Bo Jackson, *RB* | **1985** - Bo Jackson, *RB* | |
| **2010** - Cam Newton, *QB* | **2010** - Cam Newton, *QB* | |

| Davey O'Brien Award<br>*Best quarterback* | Manning Award<br>*Best quarterback* | Outland Trophy[31]<br>*Best interior lineman* | Lombardi Award[32]<br>*Best lineman/linebacker* | Jim Thorpe Award[33]<br>*Best defensive back* |
| --- | --- | --- | --- | --- |
| **2010** - Cam Newton | **2010** - Cam Newton | **1958** - Zeke Smith, *G*<br>**1988** - Tracy Rocker, *DT* | **1988** - Tracy Rocker, *DT*<br>**2010** - Nick Fairley, *DT* | **2004** - Carlos Rogers, *CB* |

### Coaches

| Paul "Bear" Bryant Award[34]<br>*Coach of the Year* | Home Depot Award[35]<br>*Coach of the Year* | Bowden Award[36]<br>*Coach of the Year* | Broyles Award[37]<br>*Best assistant coach* |
| --- | --- | --- | --- |
| **1993** - Terry Bowden | **2010** - Gene Chizik | **2010** - Gene Chizik | **2004** - Gene Chizik |
| **2004** - Tommy Tuberville | | | **2010** - Gus Malzahn |
| **2010** - Gene Chizik | | | |

## All-Americans

| Name | Position | Years | Source |
|---|---|---|---|
| Jimmy Hitchcock | HB | 1932 | WCFF |
| Walter Gilbert | C | 1933–1936 | |
| Monk Gafford | RB | 1942 | |
| Caleb "Tex" Warrington | C | 1944 | FWAA, WCFF |
| Travis Tidwell | RB | 1949 | Williamson |
| Jim Pyburn | WR | 1954 | |
| Joe Childress | RB | 1955 | FWAA |
| Frank D'Agostino | T | 1955 | AFCA |
| Fob James | RB | 1955 | INS |
| Jimmy Phillips | DE | 1957 | AFCA, FWAA, WCFF |
| Zeke Smith | OG | 1958–1959 | AFCA, FWAA, WCFF |
| Jackie Burkett | C | 1958 | AFCA |
| Ken Rice | OT | 1959–1960 | AFCA, FWAA, WCFF |
| Ed Dyas | RB | 1960 | FWAA |
| Jimmy Sidle | RB | 1963 | FWAA |
| Tucker Frederickson | RB | 1964 | FWAA, WCFF |
| Jack Thornton | DT | 1965 | NEA |
| Bill Cody | LB | | |
| Freddie Hyatt | WR | 1967 | TFN |
| David Campbell | DT | 1968 | NEA |
| Buddy McClinton | DB | 1969 | AFCA, FWAA, WCFF |
| Larry Willingham | DB | 1970 | AFCA, FWAA, WCFF |
| Pat Sullivan | QB | 1971 | AFCA, FWAA, WCFF |
| Terry Beasley | WR | 1971 | AFCA, FWAA, WCFF |
| Mike Fuller | S | 1974 | |
| Ken Bernich | LB | 1974 | AFCA, WCFF |
| Neil O'Donoghue | PK | 1976 | TFN |
| Keith Uecker | OG | 1981 | Mizlou |
| Bob Harris | SS | | |
| David King | CB | | |
| Donnie Humphrey | DT | 1983 | WTBS |
| Gregg Carr | LB | 1984 | AFCA, WCFF |
| Bo Jackson | RB | 1983–1985 | AFCA, FWAA, WCFF |
| Lewis Colbert | P | 1985 | AFCA |
| Ben Tamburello | C | 1986 | AFCA, FWAA, WCFF |
| Brent Fullwood | RB | 1986 | AFCA, FWAA, WCFF |
| Aundray Bruce | LB | 1987 | AFCA, WCFF |

| Kurt Crain | LB | 1987 | AP |
| --- | --- | --- | --- |
| Stacy Searels | OT | 1987 | AP, TFN |
| Tracy Rocker | DT | 1987–1988 | AFCA, FWAA, WCFF |
| Walter Reeves | TE | 1988 | TSN |
| Benji Roland | DT |  |  |
| Ed King | OG | 1989–1990 | AFCA, FWAA, WCFF |
| Craig Ogletree | LB | 1989 | TSN |
| David Rocker | DT | 1990 | AFCA, WCFF |
| Wayne Gandy | OT | 1993 | AP, FWAA, SH |
| Terry Daniel | P | 1993 | AFCA, FWAA, WCFF |
| Brian Robinson | SS |  |  |
| Frank Sanders | WR | 1994 | AP, FWAA, SH |
| Chris Shelling | SS | 1994 | FWAA, SH |
| Victor Riley | OT | 1997 | AFCA |
| Takeo Spikes | LB | 1997 | TSN |
| Damon Duval | PK | 2001 | AFCA, WCFF |
| Karlos Dansby | LB | 2003 | AFCA |
| Marcus McNeill | OT | 2004–2005 | AP, CBS, FWAA, SI, Rivals, CFN, WCFF |
| Carlos Rogers | CB | 2004 | AP, FWAA, WCFF |
| Junior Rosegreen | SS | 2004 | SI, CBS |
| Carnell Williams | RB | 2004 | AFCA |
| Kenny Irons | RB | 2005 | Rivals |
| Tim Duckworth | OG | 2006 | Rivals |
| Quentin Groves | DE | 2006 | Rivals |
| Ben Grubbs | OG | 2006 | Rivals, ESPN, PFW |
| David Irons | CB | 2006 | Rivals |
| Cam Newton | QB | 2010 | AFCA, AP, Rivals, SI, WCFF |
| Lee Ziemba | OT | 2010 | AFCA, FWAA, SI, WCFF |
| Nick Fairley | DT | 2010 | AP, FWAA, Rivals, SI, WCFF |

## Tigers in the NFL

Ronnie Brown was the #2 pick in the 2005 NFL Draft

There have been 245 Auburn players drafted into the National Football League, with 15 earning 30 All-Pro honors, 27 making Pro Bowl appearances, and 23 playing in the Super Bowl.

The Dow Jones College-Football Success Index ranked Auburn as the eighth best program in the nation, with the second highest Draft Value which indicate "that a school's players perform better than NFL scouts seem to expect".[38] Auburn is tied (with Miami) for second most Top 5 NFL Draft picks this decade, and The Plains have produced 25 first round draft picks overall.

## "Running Back U"

Auburn has several former running backs currently playing that position in the NFL (see below). They carry on a long legacy of top NFL backs from Auburn such as Tucker Frederickson, William Andrews, Joe Cribbs, James Brooks, Rudi Johnson, Stephen Davis, James Bostic, Brandon Jacobs, Lionel James, Brent Fullwood, Carnell Williams, Ronnie Brown, Heath Evans, Kenny Irons, Ben Tate, Fred Beasley,Kevin McCleod, Tommie Agee and Bo Jackson. Over the years 1987-2008, there have been 15 Tiger running backs drafted into the NFL, with several more successfully signing as undrafted free-agents.

Cadillac Williams evades a tackler.

## Current NFL players

There are a number of former Auburn players currently listed on NFL rosters. These players include five running backs, four linebackers, four wide receivers, six cornerbacks, two quarterbacks, one placekicker and eleven linemen including five nose tackles, two guards, two offensive tackles and two defensive ends.

| Name | Position | Team |
| --- | --- | --- |
| Darvin Adams | WR | Carolina Panthers |
| Devin Aromashodu | WR | Minnesota Vikings |
| Ronnie Brown | RB | Philadelphia Eagles |
| Jason Campbell | QB | Oakland Raiders |
| Karlos Dansby | LB | Miami Dolphins |
| King Dunlap | OT | Philadelphia Eagles |
| Nick Fairley | DT | Detroit Lions |
| Mario Fannin | RB | Denver Broncos |
| Tyronne Green | OG | San Diego Chargers |
| Quentin Groves | DE | Oakland Raiders |
| Ben Grubbs | OG | Baltimore Ravens |
| Will Herring | LB | New Orleans Saints |
| Roderick Hood | DB | St. Louis Rams |
| Spencer Johnson | DT | Buffalo Bills |
| Pat Lee | DB | Green Bay Packers |
| Senderrick Marks | DT | Tennessee Titans |
| Marcus McNeill | OT | San Diego Chargers |
| Cam Newton | QB | Carolina Panthers |
| Benjamin Obomanu | WR | Seattle Seahawks |
| Jerraud Powers | DB | Indianapolis Colts |
| Jay Ratliff | DT | Dallas Cowboys |
| Carlos Rogers | DB | San Francisco 49ers |
| Pat Sims | DT | Cincinnati Bengals |
| Takeo Spikes | LB | San Diego Chargers |
| Ben Tate | RB | Houston Texans |
| Zach Clayton | DT | Tennessee Titans |
| Reggie Torbor | LB | Buffalo Bills |
| Jonathan Wilhite | DB | Denver Broncos |
| Carnell Williams | RB | St. Louis Rams |
| Lee Ziemba | OT | Carolina Panthers |

## 2011 NFL Draft

The following former Tigers were drafted in the most recent NFL Draft:

| Name | Position | Team | Round |
|---|---|---|---|
| Cam Newton | QB | Carolina Panthers | 1 |
| Nick Fairley | DT | Detroit Lions | 1 |
| Lee Ziemba | OT | Carolina Panthers | 7 |
| Zach Clayton | DT | Tennessee Titans | 7 |

## Hall of Fame

| Name | Position | Inducted |
|---|---|---|
| Frank Gatski | C | 1985 |

## Bowl history

Auburn football teams have been invited to participate in 36 total bowls and have garnered a record of 21–13–2. Auburn ranks as one of the best programs in the nation in success in bowl games. Auburn ranks 15th in all-time bowl appearances with 36, 12th in all-time bowl wins with 21, and 9th in all-time bowl win percentage (minimum of 10 games) at .611. On January 10, 2011, Auburn defeated Oregon in the BCS National Championship Game, 22-19. Auburn has won 5 straight bowl games and 8 out of their last 9. Auburn's 5 straight bowl wins is tied for the longest current bowl game win streak.

| W/L | Date | PF | Opponent | PA | Bowl |
|---|---|---|---|---|---|
| T | 01-01-1937 | 7 | Villanova | 7 | Bacardi Bowl |
| W | 01-01-1938 | 6 | Michigan St. | 0 | Orange Bowl |
| L | 01-01-1954 | 13 | Texas Tech | 35 | Gator Bowl |
| W | 12-31-1954 | 33 | Baylor | 13 | Gator Bowl |
| L | 12-31-1955 | 13 | Vanderbilt | 25 | Gator Bowl |
| L | 01-01-1964 | 7 | Nebraska | 13 | Orange Bowl |
| L | 12-18-1965 | 7 | Mississippi | 13 | Liberty Bowl |
| W | 12-28-1968 | 34 | Arizona | 10 | Sun Bowl |
| L | 12-31-1969 | 7 | Houston | 36 | Bluebonnet Bowl |
| W | 01-02-1971 | 35 | Mississippi | 28 | Gator Bowl |
| L | 01-01-1972 | 22 | Oklahoma | 40 | Sugar Bowl |
| W | 12-30-1972 | 24 | Colorado | 3 | Gator Bowl |
| L | 12-29-1973 | 17 | Missouri | 34 | Sun Bowl |
| W | 12-30-1974 | 27 | Texas | 3 | Gator Bowl |
| W | 12-18-1982 | 33 | Boston College | 26 | Tangerine Bowl |
| W | 01-02-1984 | 9 | Michigan | 7 | Sugar Bowl |
| W | 12-27-1984 | 21 | Arkansas | 15 | Liberty Bowl |
| L | 01-01-1986 | 16 | Texas A&M | 36 | Cotton Bowl Classic |
| W | 01-01-1987 | 16 | Southern California | 7 | Florida Citrus Bowl |
| T | 01-01-1988 | 16 | Syracuse | 16 | Sugar Bowl |

| | | | | | |
|---|---|---|---|---|---|
| L | 01-02-1989 | 7 | Florida St. | 13 | Sugar Bowl |
| W | 01-01-1990 | 31 | Ohio St. | 14 | Hall of Fame Bowl |
| W | 12-29-1990 | 27 | Indiana | 23 | Peach Bowl |
| L | 01-01-1996 | 14 | Penn St. | 43 | Outback Bowl |
| W | 12-31-1996 | 32 | Army | 29 | Independence Bowl |
| W | 01-02-1998 | 21 | Clemson | 17 | Peach Bowl |
| L | 01-01-2001 | 28 | Michigan | 31 | Florida Citrus Bowl |
| L | 12-31-2001 | 10 | North Carolina | 16 | Peach Bowl |
| W | 01-01-2003 | 13 | Penn St. | 9 | Capital One Bowl |
| W | 12-31-2003 | 28 | Wisconsin | 14 | Music City Bowl |
| W | 01-03-2005 | 16 | Virginia Tech | 13 | Sugar Bowl |
| L | 01-02-2006 | 10 | Wisconsin | 24 | Capital One Bowl |
| W | 01-01-2007 | 17 | Nebraska | 14 | Cotton Bowl Classic |
| W | 12-31-2007 | 23 | Clemson | 20 | Chick-fil-A Bowl |
| W | 01-01-2010 | 38 | Northwestern | 35 | Outback Bowl |
| W | 01-10-2011 | 22 | Oregon | 19 | BCS National Championship Game |
| W | 12-31-2011 | 43 | Virginia | 24 | Chick-fil-A Bowl |

# Future schedules

## 2012 schedule

| Date | Time | Opponent[#] | Rank[#] | Site | TV | Result | Attendance |
|---|---|---|---|---|---|---|---|
| September 1 | TBD | vs. Clemson* | | Georgia Dome • Atlanta, GA (Chick-fil-A College Kickoff) | TBD | | |
| September 8 | TBD | at Mississippi State | | Davis Wade Stadium at Scott Field • Starkville, MS | TBD | | |
| September 15 | TBD | Louisiana-Monroe* | | Jordan-Hare Stadium • Auburn, AL | TBD | | |
| September 22 | TBD | LSU | | Jordan-Hare Stadium • Auburn, AL (Tiger Bowl) | TBD | | |
| October 6 | TBD | Arkansas | | Jordan-Hare Stadium • Auburn, AL | TBD | | |
| October 13 | TBD | at Ole Miss | | Vaught-Hemingway Stadium • Oxford, MS | TBD | | |

| | | | | | | | |
|---|---|---|---|---|---|---|---|
| October 20 | TBD | at Vanderbilt | | Vanderbilt Stadium • Nashville, TN | TBD | | |
| October 27 | TBD | Texas A&M | | Jordan-Hare Stadium • Auburn, AL | TBD | | |
| November 3 | TBD | New Mexico State* | | Jordan-Hare Stadium • Auburn, AL | TBD | | |
| November 10 | TBD | Georgia | | Jordan-Hare Stadium • Auburn, AL (Deep South's Oldest Rivalry) | TBD | | |
| November 17 | TBD | Alabama A&M* | | Jordan-Hare Stadium • Auburn, AL | TBD | | |
| November 24 | TBD | at Alabama | | Bryant–Denny Stadium • Tuscaloosa, AL (Iron Bowl) | TBD | | |

*Non-conference game. † Homecoming. #Rankings from AP Poll. All times are in Central Standard Time.

# References


[1]  (http://fs.ncaa.org/Docs/stats/football_records/2011/Awards.pdf)

[2]  http://www.auburntigers.com

[3]  "Division I-A All-Time Wins" (http://www.cfbdatawarehouse.com/data/misc/div_ia_wins.php). College Football Data Warehouse. 2011. . Retrieved 2011-01-21.

[4]  "I-A Winning Percentage 1986-2010 (25 years)" (http://football.stassen.com/cgi-bin/records/calc-wp.pl?start=1986&end=2010& rpct=100&ss=on&se=on&c1a=on&by=Win+Pct). Stassen College Football Information. 2011. . Retrieved 2011-01-21.

[5]  "I-A Winning Percentage 1955-2010" (http://football.stassen.com/cgi-bin/records/calc-wp.pl?start=1955&end=2010&rpct=100& ss=on&se=on&c1a=on&by=Win+Pct). Stassen College Football Information. 2011. . Retrieved 2011-01-21.

[6]  "I-A Winning Percentage 1892-2010" (http://football.stassen.com/cgi-bin/records/calc-wp.pl?start=1892&end=2010&rpct=75& ss=on&se=on&c1a=on&by=Win+Pct). Stassen College Football Information. 2011. . Retrieved 2011-01-21.

[7]  "Billingsley's All Time Top Programs" (http://www.cfrc.com/Archives/Top_Programs_2010.htm). College Football Research Center. 2011. . Retrieved 2011-01-21.

[8]  "Billingsley's Top 200 Teams of All Time" (http://www.cfrc.com/Archives/Top_200_2010.htm). College Football Research Center. 2011. . Retrieved 2011-01-21.

[9]  "College Football Prestige Rankings: Nos. 21-119" (http://sports.espn.go.com/ncf/news/story?id=3842161). 2009. . Retrieved 2010-02-14.

[10]  "Final AP Poll Appearances Summary" (http://www.collegepollarchive.com/football/ap/app_final.cfm?sort=totapp&decade=all& rows=all). AP Poll Archive. 2011. . Retrieved 2011-01-21.

[11]  "Total AP Poll Appearances Summary" (http://www.collegepollarchive.com/football/ap/app_total.cfm?sort=totapp&decade=all& rows=all). AP Poll Archive. 2011. . Retrieved 2011-01-21.

[12]  "Auburn in the Polls" (http://www.cfbdatawarehouse.com/data/div_ia/sec/auburn/in_the_polls.php). College Football Data Warehouse. 2011. . Retrieved 2011-01-21.

[13]  http://auburntigers.cstv.com/sports/m-footbl/spec-rel/041411aaa.html

[14]  "Conference Record 1992-2008 (SEC West)" (http://football.stassen.com/cgi-bin/records/conference.pl?start=1992&end=2008& team=Alabama&team=Arkansas&team=Auburn&team=LouisianaState&team=Mississippi&team=MississippiState). Stassen College Football Information. 2009. . Retrieved 2009-01-09.

[15]  "Auburn Traditions" (http://auburntigers.cstv.com/trads/aub-trads.html). Auburn University. 2006. . Retrieved 2006-09-01.

[16]  "Past Division I Football Bowl Subdivision (Division I FBS) National Champions (formerly called Division I-A)" (http://www.ncaa.org/ champadmin/ia_football_past_champs.html). .

[17]  http://sports.espn.go.com/ncf/columns/story?columnist=schlabach_mark&id=5274077

[18]  "Auburn Wins Peoples National Championship Poll" (http://auburn.scout.com/2/338643.html). Scout.com. 2005. . Retrieved 2006-07-28.

[19] "FWAA vacates USC 2004 title, doesn't anoint Auburn" (http://www.al.com/sports/index.ssf/2010/08/
fwaa_vacates_usc_2004_title_do.html). *The Birmingham News*. 2010. . Retrieved 2010-08-26.

[20] "Auburn Football 2007 Media Guide" (http://www.cstv.com/auto_pdf/p_hotos/s_chools/aub/sports/m-footbl/auto_pdf/
07-mg-history). Auburn University. 2007. . Retrieved 2007-09-30.

[21] http://www.cfbdatawarehouse.com/data/div_ia/sec/auburn/yearly_totals.php

[22] "The best Walk in America" (http://espn.go.com/page2/s/maisel/031120auburn.html). ESPN.com. 2003. . Retrieved 2007-10-13.

[23] Barnhart, Tony (2000). *Southern fried football: the history, passion, and glory of the great Southern game* (http://books.google.com/
books?id=QR8MAAAAYAAJ&q=Wreck+Tech+Pajama+Parade&dq=Wreck+Tech+Pajama+Parade&hl=en&
ei=79hhTZaiFo7TgQfloKm-Ag&sa=X&oi=book_result&ct=result&resnum=2&ved=0CD8Q6AEwAQ). Triumph. p. 49. ISBN Virginia. .

[24] 2010 Auburn Football Media Guide, p. 157

[25] "Auburn to name Chizik as coach" (http://sports.espn.go.com/ncf/news/story?id=3767115). ESPN.com. 2008-12-15. . Retrieved
2010-03-11.

[26] 2010 Auburn Football Media Guide, pp. 136–143

[27] 2010 Auburn Football Media Guide, pp. 184–193

[28] The National Collegiate Athletic Association (NCAA). "National Poll Rankings" (http://fs.ncaa.org/Docs/stats/football_records/DI/
2010/2010FBS.pdf) (PDF). *2010 NCAA Division I Football Bowl Subdivision Records*. NCAA.org. pp. 68–77. . Retrieved 2011-03-11.

[29] "Heisman Trophy Winners" (http://www.heisman.com/winners/hsmn-winners.html). heisman.com. . Retrieved 2007-12-14.

[30] Alder, James. "Walter Camp Award Winners" (http://football.about.com/cs/history/a/waltercampaward.htm). About.com. . Retrieved
2007-12-14.

[31] "All-Time Outland Trophy Winners" (http://www.sportswriters.net/fwaa/awards/outland/winners.html). Football Writers Association
of America. . Retrieved 2007-12-14.

[32] "The Rotary Lombardi Award Website — Winners" (http://www.rotarylombardiaward.org/index.php?option=com_content&
task=blogcategory&id=25&Itemid=53). Rotary Club of Houston. . Retrieved 2007-12-14.

[33] "The Jim Thorpe Award — Past Winners" (http://web.archive.org/web/20071111163652/http://jimthorpeassoc.org/Awards/
JTAPastWinners.html). The Jim Thorpe Association. Archived from the original (http://www.jimthorpeassoc.org/Awards/
JTAPastWinners.html) on 2007-11-11. . Retrieved 2007-12-14.

[34] "Paul "Bear" Bryant Previous Winners" (http://www.americanheart.org/downloadable/heart/11975759485160708 BB Previous Winners.
pdf). American Heart Association. . Retrieved 2007-12-14.

[35] "Home Depot Previous Winners" (http://espn.go.com/blog/sec/post?id=17725). Home Depot. . Retrieved 2010-12-08.

[36] "Chizik picks up another coaching honor" (http://espn.go.com/blog/sec/post/_/id/19340/chizik-picks-up-another-coaching-honor).
ESPN.com. . Retrieved 2011-03-07.

[37] "Former Winners of the Broyles Award" (http://web.archive.org/web/20071109083331/http://www.broylesaward.com/html/
former_winners.html). Rotary Club of Little Rock. Archived from the original (http://www.broylesaward.com/html/former_winners.
html) on 2007-11-09. . Retrieved 2007-12-14.

[38] "Dow Jones College-Football Success Index" (http://online.wsj.com/public/resources/documents/retro-collegefootball0608.html). The
Wall Street Journal. 2006. . Retrieved 2006-10-06.

## External links

- Official website (http://http://auburntigers.collegesports.com/sports/m-footbl/aub-m-footbl-body.html)

# Coach_(sport)

In sports, a **coach** is an individual involved in the direction, instruction and training of the operations of a sports team or of individual sportspeople.

## Staff

A coach, particularly in a major operation, is traditionally aided in his/her efforts by one or more assistant coaches known as the assistant coaching staff and team. The staff may include coordinators, strength and fitness specialists, and trainers.

Senior coach Ross Lyon addresses the St Kilda Football Club players in the Australian Football League prior to the 2009 AFL Grand Final

## Coaching in association football

In football, the duties of a coach can vary depending on the level they are coaching at and the country they are coaching in, amongst others. In youth football, the primary objective of a coach is to aid players in the development of their technical skills, with emphasis on the enjoyment and fair play of the game rather than physical or tactical development.[1] In recent decades, efforts have been made by governing bodies in various countries to overhaul their coaching structures at youth level with the aim of encouraging coaches to put player development and enjoyment ahead of winning matches.

In professional football, the role of the coach or trainer is limited to the training and development of a club's "first team" in most countries. The coach is aided by a number of assistant coaches, one of which carries the responsibility for the training and preparation of the goalkeepers. The coach is also assisted by medical staff and athletic preparators.

The medium to long term strategy of a football club, with regard to transfer policies, youth development and other sporting matters, is not the business of a coach in most footballing countries. The presence of a sporting director is designed to give the medium term development of a club the full attention of one professional, allowing the coach to focus on improving and producing performances from the players under their charge.[2] The system also provides a certain level of protection against overspending on players in search of instant success. In British football, the director of a professional football team is more commonly awarded the position of manager, a role that combines the duties of coach and sporting director.

The responsibilities of a European football (soccer) manager tend to be divided up in North American professional sports, where the teams usually have a separate general manager and head coach, although occasionally a person may fill both roles of general manager and head coach. While the first team coach in football (soccer) is usually an assistant to the manager who actually holds the real power, the American style general manager and head coach have clearly distinct areas of responsibilities. For example, a typical European football manager would have the final say on player lineups and contract negotiations, while in American sports these duties would be handled separately by the head coach and general manager, respectively.Also sports can be changed due to a coach deciding on the game.

## Coaching in baseball

Baseball, at least at the professional level in North America, is unique in that the individual who heads the coaching staff does not use the title of "head coach", but is instead called the field manager. Baseball "coaches" at that level are members of the coaching staff under the overall supervision of the manager, with each coach having a specialized role. The baseball field manager is essentially equivalent to head coaches in other American professional sports leagues; player transactions are handled by the general manager. The term "manager" used without qualification almost always refers to the field manager, while the general manager is often called the GM.

At amateur levels, the terminology is more similar to that of other sports. The person known as the "manager" in professional leagues is generally called the "head coach" in amateur leagues; this terminology is standard in U.S. college baseball.

## Coaching in the United States

All major U.S. collegiate sports have associations for their coaches to engage in professional development activities, but professional coaches tend to have less formal associations, and have never developed into a group resembling a union in the way that athletic players in many leagues have.

Many coaching contracts allow the termination of the coach with little notice and without specific cause, usually in the case of high-profile coaches with the payment of a financial settlement. U.S. collegiate coaching contracts require termination without the payment of a settlement if the coach is found to be in serious violation of named rules, usually with regard to the recruiting or retention of players in violation of amateur status.

Coaching is a very fickle profession, and a reversal of the team's fortune often finds last year's "Coach of the Year" to be seeking employment in the next.

Many coaches are former players of the sport themselves, and coaches of professional sports teams are sometimes retired players.

On some teams, the principal coach (usually referred to as the *head coach*) has little to do with the development of details such as techniques of play or placement of players on the playing surface, leaving this to assistants while concentrating on larger issues such as recruitment and organizational development.

Successful coaches often become as well or even better-known than the athletes they coach, and in recent years have come to command high salaries and have agents of their own to negotiate their contracts with the teams. Often the head coach of a well-known team has his or her own radio and television programs and becomes the primary "face" associated with the team.

## Preparation for coaching

Coaching courses and training seminars are increasingly available. One important role of coaches, especially youth coaches, is establishing safety for school-age athletes. This requires knowledge of CPR, prevention of dehydration, and following current concussion management guidelines.[3] [4]

## See also

- Head coach (American football)
- Manager (association football)
- Coach (baseball)
- Manager (baseball)
- Coach (basketball)
- Coach (ice hockey)
- Player-coach
- Player-manager

- Coaching staff
- Coaching tree
- Coach of the Year
- Personal trainer
- Personal trainer accreditation
- Sports trainer
- National Alliance for Youth Sports

## References

[1] "Phase 1 – The FUNdamental Phase" (http://www.fai.ie/index.php?option=com_content&view=article&id=100028&Itemid=291).
    Football Association of Ireland. 2009-06-12. . Retrieved 2010-12-12.
[2] "Manager or Coach?" (http://www.football-italia.net/blogs/sw48.html). Football Italia. 2008-09-05. . Retrieved 2011-01-30.
[3] Heading Off Sports Injuries, Newsweek, 5 February 2010 (http://www.newsweek.com/id/233115)
[4] Oregon Senate Bill 348 (http://special.registerguard.com/csp/cms/sites/web/updates/7502120-55/story.csp)

- Cassidy, Jones and Potrac, *Understanding Sports Coaching: The Social, Cultural and Pedagogical Foundations of Coaching Practice*, Routledge, 2008, ISBN 978-0-415-44272-5
- Jones, Hughes and Queenston, *An Introduction to Sports Coaching*, Routledge, 2007, ISBN 978-0-415-41131-8

## External links

- Sports coaching resources (http://www.coachingmill.com)
- Soccer coaching blog (http://www.soccerdrillstips.com/blog)

# Defensive_coordinator

A **defensive coordinator** typically refers to a coach on a gridiron football team who is in charge of the defense. Generally, along with his offensive counterpart, he represents the second level of command structure after the head coach. The defensive coordinator is generally in charge of managing all defensive players and assistant coaches, of developing a general defensive game plan, and of calling the plays for the defense during the game. At higher levels of football (college and professional), the defensive coordinator typically has a number of assistant coaches working under him; usually coaches primarily responsible for the various defensive positions on the team (such defensive line, linebackers, or defensive backs).

## See also

- List of active National Football League defensive coordinators
- American football strategy
- Offensive coordinator

# Iowa_State_University

| Iowa State University | |
|---|---|
| **Motto** | Science with practice |
| **Established** | 1858 |
| **Type** | Flagship state university<br>Land-grant<br>Space-grant |
| **Endowment** | US $570 million[1] |
| **President** | Steven Leath |
| **Academic staff** | 1,709 |
| **Students** | 29,887 (Fall 2011)[2] |
| **Undergraduates** | 24,343 (Fall 2011)[3] |
| **Postgraduates** | 4,958 (Fall 2011)[4] 586 veterinary[5] |
| **Location** | Ames, Iowa, United States |
| **Campus** | Urban, **unknown operator: u','** acres (**unknown operator: u','**km$^2$) |
| **Nickname** | Cyclones |
| **Colors** | |
| **Athletics** | Big 12<br>NCAA Division I |
| **Mascot** | Cy |
| **Affiliations** | American Association of Universities, Universities Research Association |
| **Website** | iastate.edu [6] |

**IOWA STATE UNIVERSITY**

**Iowa State University of Science and Technology**, more commonly known as **Iowa State University (ISU)**, is a public land-grant and space-grant research university located in Ames, Iowa, United States. Iowa State has produced astronauts, scientists, and Nobel and Pulitzer Prize winners, along with a host of other notable individuals in their respective fields. Until 1959 it was known as the Iowa State College of Agriculture and Mechanic Arts.

Founded in 1858 and coeducational from its start, ISU is classified as a Research University with very high research activity (RU/VH) by the Carnegie Foundation for the Advancement of Teaching.[7] The university is a group member of the prestigious American Association of Universities, Universities Research Association, and the Big 12 Conference.

Iowa State is a leader in agriculture, engineering, extension, and home economics, and created the nation's first state veterinary medicine school in 1879. In 1933, Iowa State established the Statistical Laboratory. It was and is the first research and consulting institute of its kind in the country.[8]

# History

## Beginnings

In 1856, the Iowa General Assembly enacted legislation to establish the State Agricultural College and Model Farm. Iowa Agricultural College and Model Farm (now Iowa State University) was officially established on March 22, 1858, by the legislature of the State of Iowa. Story County was chosen as the location on June 21, 1859, from proposals by Johnson, Kossuth, Marshall, Polk, and Story counties. The original farm of 648 acres (2.62 km$^2$) was purchased for a cost of \$5,379.[9]

Iowa was the first state in the nation to accept the provisions of the Morrill Act of 1862.[9] [10] Iowa subsequently designated Iowa State as the land-grant college on March 29, 1864.[10] [11] Iowa Agricultural College (Iowa State College of Agricultural and Mechanic Arts as of 1898), as a land grant institution, focused on the ideals that higher education should be accessible to all and that the university should teach liberal and practical subjects. These ideals are integral to the land-grant university.[12] [9]

The institution was coeducational from the first preparatory class admitted in 1868. The formal admitting of students began the following year, and the first graduating class of 1872 consisted of 24 men and two women.[9]

The Farm House, the first building on the Iowa State campus, was completed in 1861 before the campus was occupied by students or even classrooms. It became the home of the superintendent of the Model Farm and in later years, the deans of Agriculture, including Seaman Knapp and "Tama Jim" Wilson. Iowa State's first president, Adonijah Welch, briefly stayed at the Farm House and penned his inaugural speech in a second floor bedroom.[9]

The college's first farm tenants primed the land for agricultural experimentation. The Iowa Experiment Station was one of the university's prominent features. Practical courses of instruction were taught, including one designed to give a general training for the career of a farmer. Courses in mechanical, civil, electrical, and mining engineering were also part of the curriculum.

In 1870, President Welch and I. P. Robert, professor of agriculture, held three-day farmers' institutes at Cedar Falls, Council Bluffs, Washington, and Muscatine. These became the earliest institutes held off-campus by a land grant institution and were the forerunners of 20th century extension.

In 1872, the first courses were given in domestic economy (home economics, family and consumer sciences) and were taught by Mrs. Mary B. Welch, the president's wife. Iowa State became the first land grant university in the nation to offer training in domestic economy for college credit.[9]

In 1879, the "School" of Veterinary Science was organized, the first state veterinary college in the United States (although veterinary courses has been taught since the beginning of the College). This was originally a two-year course leading to a diploma. The veterinary course of study contained classes in zoology, botany, anatomy of domestic animals, veterinary obstetrics, and sanitary science.[13]

Beardshear Hall

William M. Beardshear was appointed President of Iowa State in 1891, and during his tenure, Iowa Agricultural College truly came of age. Beardshear developed new agricultural programs and was instrumental in hiring premier faculty members such Anson Marston, Louis B. Spinney, J.B. Weems, Perry G. Holden, and Maria Roberts. He also expanded the university administration, and the following buildings were added to the campus: Morrill Hall (1891); the Campanile (1899); Old Botany (now Carrie Chapman Catt Hall) (1892); and Margaret Hall (1895) which continue to stand today. In his honor, Iowa State named its central administrative building (Central Building) after Beardshear in 1925.[14] Today, Beardshear Hall holds the following offices: President, Vice-President, Treasurer, Secretary, Registrar, Provost, and student financial aid. Catt Hall is named after famed alumna Carrie Chapman Catt and is the home of the College of Liberal Arts and Sciences.

Iowa State celebrated its first VEISHEA on May 11–13, 1922. Wallace McKee (class of 1922) served as the first chairman of the Central Committee and Frank D. Paine (professor of electrical engineering) chose the name, based on the first letters of Iowa State's colleges: Veterinary Medicine, Engineering, Industrial Science, Home Economics, and Agriculture. VEISHEA has grown to become the largest student-run festival in the nation.[14]

The Statistical Laboratory was established in 1933, with George W. Snedecor, professor of mathematics, as the first director. It was and is the first research and consulting institute of its kind in the country.[15]

While attempting to develop a faster method of computation, mathematics and physics professor John Vincent Atanasoff conceptualized the basic tenets of what would become the world's first electronic digital computer, the Atanasoff-Berry Computer (ABC), during a drive to Illinois in 1937. These included the use of a binary system of arithmetic, the separation of computer and memory functions, and regenerative drum memory, among others. The 1939 prototype was constructed with graduate student Clifford Berry in the basement of the Physics Building.[16]

During World War II, Iowa State was one of 131 colleges and universities nationally that took part in the V-12 Navy College Training Program which offered students a path to a Navy commission.[17]

## Maturity as a university

On July 4, 1959, the college was officially renamed Iowa State University of Science and Technology. Official names given the university's divisions were the College of Agriculture, College of Engineering, College of Home Economics, College of Sciences and Humanities, and College of Veterinary Medicine.[18]

Iowa State's eight colleges today offer more than 100 undergraduate majors and 200 fields of study leading to graduate and professional degrees. The academic program at ISU includes a vigorous liberal arts education and some of the world's leading research in the biological and physical sciences.

Enrollment Services Building (Old Alumni Hall)

Breakthroughs at Iowa State changing the world are in the areas of human, social, economic, and environmental sustainability; new materials and processes for biomedical as well as industrial applications; nutrition, health, and wellness for humans and animals; transportation and infrastructure; food safety and security; plant and animal sciences; information and decision sciences; and renewable energies. The focus on technology has led directly to many research patents and inventions including the first binary computer (the ABC), Maytag blue cheese, the round hay baler, and many more.[19]

Located on a lush, **unknown operator: u','** acres (**unknown operator: u'strong'unknown operator: u','km$^2$**) campus, the university has grown considerably from its roots as an agricultural college and model farm to being recognized internationally today for its comprehensive research programs that are especially interdisciplinary. It continues to grow and set a new record for enrollment in the fall of 2010 with 28,685 students.[20]

# Academics

## Rankings

| University rankings (overall) | |
| --- | --- |
| **National** | |
| *Forbes*[21] | 181 |
| *U.S. News & World Report*[22] | 97 |
| *Washington Monthly*[23] | 68 |
| **Global** | |
| *ARWU*[24] | 151-200 |
| *QS*[25] | 289 |
| *Times*[26] | 184 |

ISU is ranked among the top 50 public universities in the U.S. and is known for its degree programs in science, engineering, and agriculture. Classified as a Carnegie RU/VH doctoral/research institution,[27] Iowa State receives nearly $300 million in research grants each year.

The university is one of 62 elected members of the prestigious Association of American Universities, an organization composed of the most highly ranked public and private research universities in the U.S. and Canada.

Overall, ISU ranks #94 in the *U.S. News & World Report* ranking of national universities[28] and #21 in the *Washington Monthly* rankings. In engineering specialties, at schools whose highest degree is a doctorate, Iowa State's agricultural engineering program is ranked third among top programs in the U.S. Aerospace engineering ranks 13th among public universities (18th overall). Chemical engineering and civil engineering both are ranked 13th among public universities (20th overall). Materials engineering is ranked 11th among public universities (17th overall). The electrical engineering program is ranked 25th and computer engineering program ranked 23rd out of all public programs.[29] The programs ranked 41st and 39th, respectively, out of all public and private programs in the country. Computer Science ranked 37th among public programs. Overall, Iowa State's engineering program ranks 37th. In 2009, ISU ranked 12th for top architecture programs in the nation and as obtained these rankings for years. ISU's Greenlee School of Journalism and Communication is notable for being among the first group of accredited journalism and mass communication programs[30] and is cited as one of the leading JMC research programs in the nation, ranked 23rd in a publication by the AEJMC.[31]

The National Science Foundation ranks ISU #94 in the nation in research and development expenditures for science and engineering and #78 in total research and development expenditures. Currently, ISU ranks #2 in license and options executed on its intellectual property and #5 in license and options that yield income.

## Parks Library

W. Robert and Ellen Sorge Parks Library

The W. Robert and Ellen Sorge Parks Library contains over 2.6 million books and subscribes to more than 98,600 journal titles, making ISU's library one of the 100 largest university libraries in the country. Named for W. Robert Parks (1915–2003), the 11th president of Iowa State University, and his wife, Ellen Sorge Parks, the original library was built in 1925 with three subsequent additions made in 1961, 1969, and 1983. The library was dedicated and named after W. Robert and Ellen Sorge Parks in 1984.[32]

Parks Library provides extensive research collections, services and information literacy instruction/information for all students. Facilities consist of the main Parks Library, the e-Library, the Veterinary Medical Library, two subject-oriented reading rooms (design and mathematics), and a remote library storage building.

The Library's extensive collections include electronic and print resources that support research and study for all undergraduate and graduate programs. Nationally recognized collections support the basic and applied fields of biological and physical sciences. The Parks Library includes four public service desks: the Learning Connections Center, the Circulation Desk, the Media Center (including Maps, Media, Microforms, and Course Reserve collections), and Special Collections. The Library's instruction program includes a required undergraduate information literacy course as well as a wide variety of subject-based seminars on effective use of Library resources for undergraduate and graduate students.

The e-Library, accessed through the Internet, provides access to local and Web-based resources including electronic journals and books, local collections, online indexes, electronic course reserves and guides, and a broad range of subject research guides.

Surrounding the first floor lobby staircase in Parks Library are eight mural panels designed by Iowa artist Grant Wood. As with *Breaking the Prairie Sod*, Wood's other Iowa State University mural painted two years later, Wood borrowed his theme for *When Tillage Begins Other Arts Follow* from a speech on agriculture delivered by Daniel Webster in 1840 at the State House in Boston. Webster said, "When tillage begins, other arts follow. The farmers therefore are the founders of human civilization." Wood had planned to create seventeen mural panels for the library, but only the eleven devoted to agriculture and the practical arts were completed. The final six, which would have hung in the main reading room (now the Periodical Room) and were to have depicted the fine arts, were never begun.[33]

## Distinctions

### Birthplace of first electronic digital computer

Iowa State is the birthplace of the first electronic digital computer, starting the world's computer technology revolution. Invented by mathematics and physics professor John Atanasoff and engineering graduate student Clifford Berry during 1937-42, the Atanasoff-Berry Computer, or ABC, pioneered important elements of modern computing, including binary arithmetic, regenerative memory, parallel processing, electronic switching elements, and separation of memory and computer functions.[34]

On October 19, 1973, U.S. Federal Judge Earl R. Larson signed his decision following a lengthy court trial which declared the ENIAC patent of Mauchly and Eckert invalid and named Atanasoff the inventor of the electronic digital computer—the Atanasoff-Berry Computer or the ABC.[35]

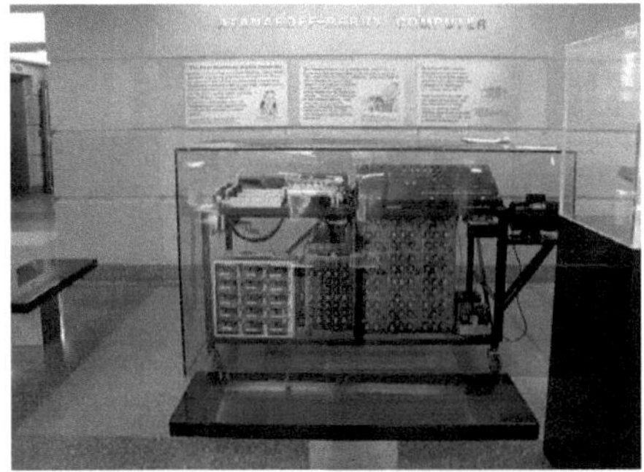

Atanasoff–Berry Computer replica on 1st floor of Durham Center, Iowa State University.

An ABC Team consisting of Ames Laboratory and Iowa State engineers, technicians, researchers and students unveiled a working replica of the Atanasoff-Berry Computer in 1997 which can be seen on display on campus in the Durham Computation Center.[36]

### Birth of cooperative extension

The Extension Service traces its roots to farmers' institutes developed at Iowa State in the late 19th century. Committed to community, Iowa State pioneered the outreach mission of being a land-grant college through creation of the first Extension Service in 1902. In 1906, the Iowa Legislature enacted the Agricultural Extension Act making funds available for demonstration projects. It is believed this was the first specific legislation establishing state extension work, for which Iowa State assumed responsibility. The national extension program was created in 1914 based heavily on the Iowa State model.[37] [38] [39]

### Manhattan Project

ISU is the only university nationwide that has a U.S. Department of Energy research laboratory physically located on its campus. Iowa State played a critical role in the development of the atomic bomb during World War II. As part of the Manhattan Project, the process to produce large quantities of high-purity uranium metal was developed at Iowa State. Iowa State provided one-third of the uranium metal used in the world's first controlled nuclear chain reaction, or atomic bomb.[40] The Ames Laboratory now focuses on more peaceful applications of materials research, usually related to increasing energy efficiency.

# Colleges

ISU is organized into eight colleges that offer 96 Bachelors degree programs, 115 Masters programs, 83 Ph.D programs, and one professional degree program in Veterinary Medicine.

ISU consists of the following colleges:

- Agriculture and Life Sciences
- Business
- Design
- Engineering
- Human Sciences
- Liberal Arts & Sciences
- Veterinary Medicine

In addition to these seven colleges, the Graduate College oversees graduate study in all fields.

# Campus and Landmarks

## Recognition

Iowa State's campus contains over 160 buildings. Several buildings, as well as the Marston Water Tower, are listed on the National Register of Historic Places.[41] The central campus includes 490 acres (2.0 km$^2$) of trees, plants, and classically designed buildings. The landscape's most dominant feature is the 20-acre (81000 m$^2$) central lawn, which was listed as a "medallion site" by the American Society of Landscape Architects in 1999, one of only three central campuses designated as such. The other two were Yale University and the University of Virginia.

The medallion located in Central Campus, immediately to the west of Curtiss Hall

Thomas Gaines, in *The Campus As a Work of Art*, proclaimed the Iowa State campus to be one of the twenty-five most beautiful campuses in the country. Gaines noted Iowa State's park-like expanse of central campus, and the use of trees and shrubbery to draw together ISU's varied building architecture. Over decades, campus buildings, including the Campanile, Beardshear Hall, and Curtiss Hall, circled and preserved the central lawn, creating a space where students study, relax, and socialize.[42]

## Campanile

The campanile was constructed during 1897-1898 as a memorial to Margaret MacDonald Stanton, Iowa State's first dean of women, who died on July 25, 1895. The tower is located on ISU's central campus, just north of the Memorial Union. The site was selected by Margaret's husband, Edgar W. Stanton, with the help of then-university president William M. Beardshear. The campanile stands 110 feet (34 m) tall on a 16 by 16 foot (5 by 5 m) base, and cost $6,510.20 to construct.[43]

The campanile is widely seen as one of the major symbols of Iowa State University. It is featured prominently on the university's official ring[44] and the university's mace,[45] and is also the subject of the university's alma mater, *The Bells of Iowa State*.[43]

The campanile as seen from the north/northwest

## Lake LaVerne

Named for Dr. LaVerne W. Noyes, who also donated the funds to see that Alumni Hall could be completed after sitting unfinished and unused from 1905 to 1907. Dr. Noyes is an 1872 alumnus. Lake LaVerne is located west of the Memorial Union and south of Alumni Hall, Carver Hall, and Music Hall. The lake was a gift from Dr. Noyes in 1916.

Lake LaVerne is the home of two mute swans named Sir Lancelot and Elaine, donated to Iowa State by VEISHEA 1935.[46] In 1944, 1970, and 1971 cygnets (baby swans) made their home on Lake LaVerne. Previously Sir Lancelot and Elaine were trumpeter swans but were too aggressive and in 1999 were replaced with two mute swans.

In early spring 2003, Lake LaVerne welcomed its newest and most current mute swan duo. In support of Iowa Department of Natural Resources efforts to re-establish the trumpeter swans in Iowa, university officials avoided bringing breeding pairs of male and female mute swans to Iowa State which means the current Sir Lancelot and Elaine are both female.[47]

## Reiman Gardens

Iowa State has maintained a horticulture garden since 1914. Reiman Gardens is the third location for these gardens. Today's gardens began in 1993 with a gift from Bobbi and Roy Reiman. Construction began in 1994 and the Gardens' initial 5 acres (20,000 m2) were officially dedicated on September 16, 1995.

Reiman Gardens has since grown to become a 14 acres (57000 $m^2$) site consisting of a dozen distinct garden areas, an indoor conservatory and an indoor butterfly "wing", butterfly emergence cases, a gift shop, and several supporting greenhouses. Located immediately south of Jack Trice Stadium on the ISU campus,

Lake Helen at Reiman Gardens

Reiman Gardens is a year-round facility that has become one of the most visited attractions in central Iowa.

The Gardens has received a number of national, state, and local awards since its opening, and its rose gardens are particularly noteworthy. It was honored with the President's Award in 2000 by All American Rose Selections, Inc., which is presented to one public garden in the United States each year for superior rose maintenance and display: "For contributing to the public interest in rose growing through its efforts in maintaining an outstanding public rose garden."[48]

## University Museums

The University Museums consist of the Brunnier Art Museum, Farm House Museum, the Art on Campus Program, the Christian Petersen Art Museum, and the Elizabeth and Byron Anderson Sculpture Garden. The Museums include a multitude of unique exhibits, each promoting the understanding and delight of the visual arts as well as attempt to incorporate a vast interaction between the arts, sciences, and technology.[49]

### Brunnier Art Museum

The Brunnier Art Museum, Iowa's only accredited museum emphasizing a decorative arts collection, is one of the nation's few museums located within a performing arts and conference complex, the Iowa State Center.[50] Founded in 1975, the museum is named after its benefactors, Iowa State alumnus Henry J. Brunnier and his wife Ann. The decorative arts collection they donated, called the Brunnier Collection, is extensive, consisting of ceramics, glass, dolls, ivory, jade, and enameled metals.

Other fine and decorative art objects from the University Art Collection include prints, paintings, sculptures, textiles, carpets, wood objects, lacquered pieces, silver, and furniture. About eight to 12 annual changing exhibitions and permanent collection exhibitions provide educational opportunities for all ages, from learning the history of a quilt hand-stitched over 100 years ago to discovering how scientists analyze the physical properties of artists' materials, such as glass or stone. Lectures, receptions, conferences, university classes, panel discussions, gallery walks, and gallery talks are presented to assist with further interpretation of objects.

### Farm House Museum

The Farm House Museum

Located near the center of the Iowa State campus, the Farm House Museum sits as a monument to early Iowa State history and culture as well as a National Historic Landmark. As the first building on campus, the Farm House was built in 1860 before campus was occupied by students or even classrooms. The college's first farm tenants primed the land for agricultural experimentation. This early practice lead to Iowa State Agricultural College and Model Farm opening its doors to Iowa students for free in 1869 under the Morrill Act (or Land-grant Act) of 1862.[51]

Many prominent figures have made the Farm House their home throughout its 150 years of use. The first president of the College, Adonijah Welch, briefly stayed at the Farm House and even wrote his inaugural speech in a bedroom on the second floor. James "Tama Jim" Wilson resided for much of the 1890s with his family at the Farm House until he joined President William McKinley's cabinet as U.S. Secretary of Agriculture. Agriculture Dean Charles Curtiss and his young family replaced Wilson and became the longest resident of Farm House.

In 1976, over 110 years after the initial construction, the Farm House became a museum after much time and effort was put into restoring the early beauty of the modest farm home. Today, faculty, students, and community members can enjoy the museum while honoring its significance in shaping a nationally recognized land-grant university. Its collection boasts a large collection of 19th and early 20th century decorative arts, furnishings and material culture reflecting Iowa State and Iowa heritage. Objects include furnishings from Carrie Chapman Catt and Charles Curtiss, a wide variety of quilts, a modest collection of textiles and apparel, and various china and glassware items.

As with many sites on the Iowa State University Campus, The Farm House Museum has a few old myths and legends associated with it. There are rumors of a ghost changing silverware and dinnerware, unexplained rattling furniture, and curtains that have opened seemingly by themselves.

The Farm House Museum is a unique on-campus educational resource providing a changing environment of exhibitions among the historical permanent collection objects that are on display. A walk through the Farm House Museum immerses visitors in the Victorian era (1860-1910) as well as exhibits colorful Iowa and local Ames history.

## Art on Campus Collection

Iowa State is home to one of the largest campus public art programs in the United States. Over 2,000 works of public art, including 600 by significant national and international artists, are located across campus in buildings, courtyards, open spaces and offices.[52]

The traditional public art program began during the Depression in the 1930s when Iowa State College's President Raymond Hughes envisioned that "the arts would enrich and provide substantial intellectual exploration into our college curricula." Hughes invited Grant Wood to create the Library's agricultural murals that speak to the founding of Iowa and Iowa State College and Model Farm. He also offered Christian Petersen a one-semester sculptor residency to design and build the fountain and bas relief at the Dairy Industry Building. In 1955, 21 years later, Petersen retired having created 12 major sculptures for the campus and hundreds of small studio sculptures.

The Art on Campus Collection is a campus-wide resource of over 2000 public works of art. Programs, receptions, dedications, university classes, Wednesday Walks, and educational tours are presented on a regular basis to enhance visual literacy and aesthetic appreciation of this diverse collection.

## Christian Petersen Art Museum

The Christian Petersen Art Museum in Morrill Hall is named for the nation's first permanent campus artist-in-residence, Christian Petersen, who sculpted and taught at Iowa State from 1934 through 1955, and is considered the founding artist of the Art on Campus Collection.

Named for Justin Smith Morrill who created the Morrill Land-Grant Colleges Act, Morrill Hall was completed in 1891. Originally constructed to fill the capacity of a library, museum, and chapel, its original uses are engraved in the exterior stonework on the east side. The building was vacated in 1996 when it was determined unsafe and was also listed in the National Register of Historic Places the same year. In 2005, $9 million was raised to renovate the building and convert it into a museum. Completed and reopened in March 2007, Morrill Hall is home to the Christian Petersen Art Museum.

As part of University Museums, the Christian Petersen Art Museum at Morrill Hall is the home of the Christian Petersen Art Collection, the Art on Campus Program, the University Museums'

Morrill Hall, home to Christian Petersen Art Museum

Visual Literacy and Learning Program, and Contemporary Changing Art Exhibitions Program. Located within the Christian Petersen Art Museum are the Lyle and Nancy Campbell Art Gallery, the Roy and Bobbi Reiman Public Art Studio Gallery, the Margaret Davidson Center for the Study of the Art on Campus Collection, the Edith D. and

Torsten E. Lagerstrom Loaned Collections Center, and the Neva M. Petersen Visual Learning Gallery. University Museums shares the James R. and Barbara R. Palmer Small Objects Classroom in Morrill Hall.[53]

### Anderson Sculpture Garden

The Elizabeth and Byron Anderson Sculpture Garden is located by the Christian Petersen Art Museum at historic Morrill Hall. The sculpture garden design incorporates sculptures, a gathering arena, and sidewalks and pathways. Planted with perennials, ground cover, shrubs, and flowering trees, the landscape design provides a distinctive setting for important works of 20th and 21st century sculpture, primarily American. Ranging from forty-four inches to nearly nine feet high and from bronze to other metals, these works of art represent the richly diverse character of modern and contemporary sculpture.[54]

The sculpture garden is adjacent to Iowa State's 22 acres (89000 m$^2$) central campus. Adonijah Welch, ISU's first president, envisioned a picturesque campus with a winding road encircling the college's majestic buildings, vast lawns of green grass, many varieties of trees sprinkled throughout to provide shade, and shrubbery and flowers for fragrance. Today, the central lawn continues to be an iconic place for all Iowa Staters, and enjoys national acclaim as one of the most beautiful campuses in the country. The new Elizabeth and Byron Anderson Sculpture Garden further enhances the beauty of Iowa State.

## Sustainability

Iowa State's composting facility "can handle more than 10,000 tons of organic wastes annually."[55] A new website tracks energy use of campus buildings[56] and the school's new $3 million dollar revolving loan fund loans money for energy efficiency and conservation projects on campus.[57] In the 2011 College Sustainability Report Card issued by the Sustainable Endowments Institute, the university received a B grade.[58]

# Student life

## Residence halls

Iowa State operates 19 on-campus residence halls. The residence halls are divided into geographical areas.

The Union Drive Association (UDA) consists of four residence halls located on the west side of campus, including Friley Hall, which has been declared one of the largest residence halls in the country.[59]

The Richardson Court Association (RCA) consists of 12 residence halls on the east side of campus.

The Towers Residence Association (TRA) are located south of the main campus. Two of the four towers, Knapp and Storms Halls, were imploded in 2005; however, Wallace and Wilson Halls still stand.

View looking east towards Roberts Hall.

Buchanan Hall is an upper-division hall housing graduate students that is nominally considered part of the RCA, despite its distance from the other buildings.

ISU also operates two apartment complexes for upperclassmen, Frederiksen Court and SUV Apartments.

| Union Drive | Richardson Court | | Towers | Apartments | Other |
|---|---|---|---|---|---|
| • Friley Hall | • Birch-Welch-Roberts Halls | • Maple Hall | • Wallace Hall | • Frederiksen Court | • Buchanan Hall |
| • Helser Hall | • Barton Hall | • Willow Hall | • Wilson Hall | • Schilleter and University Village | |
| • Martin Hall | • Lyon Hall | • Larch Hall | | | |
| • Eaton Hall | • Freeman Hall | | | | |
| | • Linden Hall | | | | |
| | • Oak-Elm Halls | | | | |

## Student government

The governing body for ISU students is the Government of Student Body or GSB. The GSB is composed of a president, vice president, finance director, cabinet appointed by the president, a clerk appointed by the vice president, senators representing each college and residence area at the university, a nine-member judicial branch and an election commission.[60]

## Student organizations

ISU has over 800 student organizations on campus that represent a variety of interests. Organizations are supported by Iowa State's Student Activities Center. Many student organization offices are housed in the Memorial Union.

Memorial Union

The Memorial Union at Iowa State University opened in September 1928 and is currently home to a number of University departments and student organizations, a bowling alley, the University Book Store, and the Hotel Memorial Union.

The original building was designed by architect, William T. Proudfoot. The building employs a classical style of architecture reflecting Greek and Roman influences. The building's design specifically complements the designs of the major buildings surrounding the University's Central Campus area, Beardshear Hall to the west, Curtiss Hall to the east, and MacKay Hall to the north. The style utilizes columns with Corinthian capitals, Paladian windows, triangular pediments, and formally balanced facades.[61]

Designed to be a living memorial for ISU students lost in World War I, the building includes a solemn memorial hall, named the Gold Star Room, which honors the names of the dead World War I, World War II, Vietnam, Korean, and War on Terrorism veterans engraved in marble. Symbolically, the hall was built directly over a library (the Browsing Library) and a small chapel, the symbol being that no country would ever send its young men to die in a war for a noble cause without a solid foundation on both education (the library) and religion (the chapel).[62]

Renovations and additions have continued through the years to include: elevators, bowling lanes, a parking ramp, a book store, food court, and additional wings.

## Greek community

ISU is home to an active Greek community. There are 50 chapters that involve 11 percent of undergraduate students. Collectively, fraternity and sorority members have raised over $82,000 for philanthropies and committed 31,416 hours to community service. In 2006, the ISU Greek community was named the best large Greek community in the Midwest.[63]

The ISU Greek Community has received multiple Jellison and Sutherland Awards from Association for Fraternal Leadership and Values, formerly the Mid-American Greek Council Association. These awards recognize the top Greek Communities in the Midwest.

| Collegiate Panhellenic Council | Interfraternity Council | | National Pan-Hellenic Council | Multicultural Greek Council |
|---|---|---|---|---|
| • Alpha Chi Omega[64] | • Acacia | • Phi Gamma Delta | • Alpha Phi Alpha | • Delta Lambda Phi |
| • Alpha Delta Pi[65] | • Adelante | • Phi Kappa Theta | • Kappa Alpha Psi | • Lambda Theta Nu |
| • Alpha Gamma Delta[66] | • Alpha Gamma Rho | • Phi Kappa Psi | • Delta Sigma Theta | • Phi Iota Alpha |
| • Alpha Omicron Pi | • Alpha Kappa Lambda | • Pi Kappa Alpha | • Omega Psi Phi | • Sigma Lambda Beta |
| • Alpha Sigma Kappa | • Alpha Sigma Phi | • Pi Kappa Phi | • Phi Beta Sigma | • Sigma Lambda Gamma |
| • Chi Omega | • Alpha Tau Omega | • Sigma Alpha Epsilon | • Zeta Phi Beta | |
| • Delta Delta Delta | • Beta Sigma Psi | • Sigma Chi | | |
| • Delta Zeta | • Beta Theta Pi | • Sigma Nu | | |
| • Gamma Phi Beta | • Delta Tau Delta | • Sigma Phi Epsilon | | |
| • Kappa Alpha Theta | • Delta Upsilon | • Sigma Pi | | |
| • Kappa Delta | • FarmHouse | • Tau Kappa Epsilon | | |
| • Kappa Kappa Gamma | • Kappa Sigma | • Theta Chi | | |
| • Phi Beta Chi | • Lambda Chi Alpha | • Theta Delta Chi | | |
| • Pi Beta Phi | • Phi Delta Theta | • Theta Xi | | |
| • Sigma Kappa | | • Triangle | | |

The first fraternity, Delta Tau Delta, was established at Iowa State in 1875, six years after the first graduating class entered Iowa State. The first sorority, I.C. Sorocis, was established only two years later, in 1877. I.C. Sorocis later became a chapter of the first national sorority at Iowa State, Pi Beta Phi. Anti-Greek rioting occurred in 1888. As reported in *The Des Moines Register*, "The anti-secret society men of the college met in a mob last night about 11 o'clock in front of the society rooms in chemical and physical hall, determined to break up a joint meeting of three secret societies." In 1891, President William Beardshear banned students from joining secret college fraternities, resulting in the eventually closing of all formerly established fraternities. President Storms lifted the ban in 1904.[67]

Following the lifting of the fraternity ban, the first twelve national fraternities (IFC) installed on the Iowa State campus between 1904 and 1913 were, in order, Sigma Nu, Sigma Alpha Epsilon, Phi Gamma Delta, Alpha Tau Omega, Kappa Sigma, Theta Xi, Acacia, Phi Sigma Kappa, Delta Tau Delta, Pi Kappa Alpha, and Phi Delta Theta.[68] Though some have suspended their chapters at various times, ten of the original twelve fraternities are active in 2008. Many of these chapters existed on campus as local fraternities before being reorganized as national fraternities, prior to 1904.

## School newspaper

The *Iowa State Daily* is the university's student newspaper and is the nation's largest student run newspaper. The *Daily* has its roots from a news sheet titled the *Clipper*, which was started in the spring of 1890 by a group of students at Iowa Agricultural College led by F.E. Davidson. The *Clipper* soon led to the creation of the *Iowa Agricultural College Student*, and the beginnings of what would one day become the *Iowa State Daily*.

## Campus radio

88.5 KURE is the university's student-run radio station. Programming for KURE includes ISU sports coverage, talk shows, the annual quiz contest Kaleidoquiz, and various music genres.

# Athletics

The "Cyclones" name dates back to 1895. That year, Iowa suffered an unusually high number of devastating cyclones (as tornadoes were called at the time). In September, the Iowa State football team traveled to Northwestern University and defeated its highly-regarded team by a score of 36-0. The next day, the *Chicago Tribune*'s headline read "Struck by a Cyclone: It Comes from Iowa and Devastates Evanston Town."[69] The article reported that "Northwestern might as well have tried to play football with an Iowa cyclone as with the Iowa team it met yesterday." The nickname stuck and the Iowa State team had made a name for itself.

The school colors are cardinal and gold. The mascot is Cy the Cardinal, introduced in 1954. Since a cyclone was determined to be difficult to depict in costume, the cardinal was chosen in reference to the school colors. A contest was held to select a name for the mascot, with the name Cy being chosen as the winner. In early Summer 2007, Cy was voted by fans on the CBS Sports website as the "Most Dominant College Mascot on Earth".[70] In 2009, Cy won the Capital One Mascot competition.

The Iowa State Cyclones are a member of the Big 12 Conference and compete in NCAA Division I-A, fielding 16 varsity teams in 12 sports. The Cyclones also compete in and are a founding member of the Central States Collegiate Hockey League of the American Collegiate Hockey Association. Iowa State teams and individuals have achieved great success, including national championships in wrestling, gymnastics, cross country, and club ice hockey.

Iowa State's intrastate archrival is the University of Iowa whom it competes annually for the Iowa Corn Cy-Hawk Series trophy, an annual athletic competition between the two schools. Sponsored by the Iowa Corn Growers Association, the competition includes all head-to-head regular season competitions between the two rival universities in all sports.

## Football

ISU marching band providing pre-game entertainment at Jack Trice Stadium.

Football first made its way onto the Iowa State campus in 1878 as a recreational sport, but it was not until 1892 that Iowa State organized its first team to represent the school in football. In 1894, college president William M. Beardshear spearheaded the foundation of an athletic association to officially sanction Iowa State football teams. The 1894 team finished with a 6-1 mark, including a 16-8 victory over what is now the University of Iowa.[71] The Cyclones compete each year for traveling trophies. Since 1977, Iowa State and Iowa compete annually for the Cy-Hawk Trophy. Iowa State competes with conference rivals Missouri for the Telephone Trophy and Kansas State in the annual Farmageddon series.

The Cyclones play its home games at Jack Trice Stadium, named after Jack Trice, ISU's first African-American athlete and also the first and only Iowa State athlete to die from injuries sustained during athletic competition. Trice died three days after his first game playing for Iowa State against Minnesota in Minneapolis on October 6, 1923. Suffering from a broken collarbone early in the game, he continued to play until he was trampled by a group of Minnesota players. It is disputed whether he was trampled purposely or if it was by accident. The stadium was named in his honor in 1997 and is the only NCAA Division I-A stadium named after an African-American.[72] Jack Trice Stadium, formerly known as Cyclone Stadium, opened on September 20, 1975, with a win against the Air Force Academy.

Beginning with its first appearance in the 1971 Sun Bowl, the Cyclones have earned ten trips to bowl games. In its most recent bowl, Iowa State defeated Minnesota in the 2009 Insight Bowl, 14-13, in Tempe, Arizona. Head coach Paul Rhoads, hired in 2009, has led the cyclone football team to signature wins in each of his 1st three years. Their upset win at Nebraska in 2009 led to a now famous post-game locker room video where coach Rhoads emotionally yells "I am so proud to be your coach!" as his team erupts. The YouTube video went viral and the program followed with their 1st win ever over the Texas Longhorns in Austin in 2010. Arguably the biggest win in Cyclone football history occurred on Friday, November 18, 2011 at Jack Trice Stadium during a nationally televised Friday night game on ESPN. Iowa State shocked the country with a stunning come-from-behind 37-31 victory in double overtime over #2 ranked, and undefeated Oklahoma State. Spoiling the cowboys chance at a national championship. Again, Rhoads and team were viewed in an emotional post-game locker room video. The team has adopted the motto "all in" and are gaining national attention by sports talk shows, media and recruits nationwide.

## Men's Basketball

Hopes of "Hilton Magic" returning took a boost with the hiring of ISU alum, Ames native, and fan favorite Fred Hoiberg as coach of the men's basketball team in April 2010. Hoiberg ("The Mayor") played three seasons under legendary coach Johnny Orr and one season under future Chicago Bulls coach Tim Floyd during his standout collegiate career as a Cyclone (1991–95). Orr laid the foundation of success in men's basketball upon his arrival from Michigan in 1980 and is credited with building Hilton Magic. Besides Hoiberg, other Cyclone greats played for Orr and brought winning seasons, including Jeff Grayer, Barry Stevens, and walk-on Jeff Hornacek. The 1985-86 Cyclones were one of the most memorable. Orr coached the team to second place in the Big Eight and produced one of his greatest career wins, a victory over his former team and No. 2 seed Michigan in the second round of the NCAA tournament.

Under coaches Floyd (1995–98) and Larry Eustachy (1998–2003), Iowa State achieved even greater success. Floyd took the Cyclones to the Sweet Sixteen in 1997 and Eustachy led ISU to two consecutive Big 12 regular season conference titles in 1999-2000 and 2000–01, plus the conference tournament title in 2000. Seeded No. 2 in the 2000 NCAA tournament, Eustachy and the Cyclones defeated UCLA in the Sweet Sixteen before falling to Michigan State, the eventual NCAA Champion, in the regional finals by a score of 75-64 (the differential representing the Spartans' narrowest margin of victory in the tournament). Standout Marcus Fizer and Jamaal Tinsley were scoring leaders for the Cyclones who finished the season 32-5. Tinsley returned to lead the Cyclones the following year with another conference title and No. 2 seed, but ISU finished the season with a 25-6 overall record after a stunning loss to No. 15 seed Hampton in the first round.

Of Iowa State's 13 NCAA Tournament appearances, the Cyclones have reached the Sweet Sixteen four times (1944, 1986, 1997, 2000), made two appearances in the Elite Eight (1944, 2000), and reached the Final Four once in 1944.[73]

## Women's Basketball

Iowa State is known for having one of the most successful women's basketball programs in the nation. Since the founding of the Big 12, Coach Bill Fennelly and the Cyclones have won three conference titles (one regular season, two tournament), and have advanced to the Sweet Sixteen five times (1999–2001, 2009, 2010) and the Elite Eight twice (1999, 2009) in the NCAA Tournament. The team is also a leader in attendance, finishing third in the nation in both 2009 and 2010.[74]

## Wrestling

The storied wrestling program has captured the NCAA wrestling tournament title eight times between 1928 and 1987,[75] and won the Big 12 Conference Tournament three consecutive years, 2007-2009. On February 7, 2010, the Cyclones became the first collegiate wrestling program to record its 1,000th dual win in program history by defeating Arizona State, 30-10, in Tempe, Arizona.

In 2002, under Coach Bobby Douglas, Iowa State became the first school to produce a four-time, undefeated NCAA champion in Cael Sanderson, who also took the gold medal at the 2004 Olympic Games in Athens, Greece. Dan Gable, another legendary ISU wrestler, is famous for having lost only one match in his entire Iowa State collegiate career - his last, and winning gold at the 1972 Olympics in Munich, Germany, while not giving up a single point.

In 2013, Iowa State will host its eighth NCAA Wrestling Championships. The Cyclones hosted the first NCAA championships in 1928.

## Volleyball

In volleyball, Coach Christy Johnson-Lynch has led the Cyclones to four NCAA tournament appearances since her Iowa State arrival in 2005, including two Sweet Sixteen appearances and an Elite Eight. In 2009, Iowa State finished the season second in the Big 12 behind Texas with a 27-5 record and ranked #6, its highest ever finish.

# VEISHEA celebration

Iowa State is widely known for VEISHEA, an annual education and entertainment festival held on campus each spring. The name VEISHEA is derived from the initials of ISU's five original colleges, forming an acronym as the university existed when the festival was founded in 1922:

- Veterinary Medicine
- Engineering
- Industrial Science
- Home Economics
- Agriculture

VEISHEA is the largest student run festival in the nation, bringing in tens of thousands of visitors to the

The VEISHEA 2006 Battle of the Bands.

campus each year. The celebration features an annual parade and many open-house demonstrations of the university facilities and departments. Campus organizations exhibit products, technologies, and hold fund raisers for various charity groups. In addition, VEISHEA brings speakers, lecturers, and entertainers to Iowa State, and throughout its over eight decade history, it has hosted such distinguished guests as Bob Hope, John Wayne, Presidents Harry Truman, Ronald Reagan, and Lyndon Johnson, and performers Diana Ross, Billy Joel, Sonny and Cher, The Who, The Goo Goo Dolls, Bobby V, and The Black Eyed Peas.[76]

The 2007 VEISHEA festivities marked the start of Iowa State's year-long sesquicentennial celebration.

## Notable people

As with any major public university, many Iowa State University alumni have achieved fame or notoriety after graduating. These people include astronauts, scientists, Nobel laureates, Pulitzer Prize winners, statesmen, academicians, CEOs, entrepreneurs, athletes, film and television actors, and a host of other notable individuals in their respective fields. USDA buildings and their architectural structures in Washington, D.C. bear more names of Iowa State alumni than those from any other university.[77] More than one-third of the Fortune 500 companies have Iowa State alumni in leadership positions.[77]

## Iowa State chronology

Events occurring in the same year did not necessarily happen in the order presented here.

Carrie Chapman Catt, suffragist, early feminist, political activist, League of Women Voters founder, and Iowa State alumna (1880).

George Washington Carver was a student and faculty member at Iowa State.

| Year | Event |
| --- | --- |
| 1856 | Iowa General Assembly enacts legislation for creation of the State Agricultural College and Model Farm |
| 1859 | Story County is the chosen county for the State Agricultural College and Model Farm |
| 1860 | Construction starts on the first building on campus, Farm House |
| 1862 | Morrill Act of 1862 is passed; college to be named Iowa State Agricultural College |
| 1869 | First graduating class enters Iowa State[78] |
| 1875 | The first national fraternity, Delta Tau Delta, opens at Iowa State |
| 1876 | The university cemetery is opened. One of the very few active cemeteries associated with a university campus in the U.S.[79] |
| 1877 | The first national sorority, Pi Beta Phi, opens at Iowa State |
| 1879 | The School of Veterinary Science is formally organized. It's the first of its kind in the United States.[80] |
| 1890 | Student newspaper *Iowa Agricultural College Student* is founded. Later to be named the Iowa State Daily |
| 1895 | Football team nicknamed Cyclones for their performance against Northwestern University |
| 1898 | The college is divided into "divisions": Agriculture, Engineering, Science and Philosophy, and Veterinary Medicine |
| 1898 | Renamed the Iowa State College of Agriculture and Mechanic Arts[81] |
| 1905 | First Agricultural Engineering program in the world established |
| 1913 | The college roads are paved |
| 1922 | VEISHEA is established[82] |
| 1923 | Jack Trice is mortally injured during a football game against Minnesota |
| 1933 | First statistics laboratory in the U.S. is established[83] |
| 1939 | The Atanasoff–Berry Computer (ABC) is invented. The Atanasoff-Berry Computer was the world's first electronic digital computer.[84][85] |
| 1945 | Campus production reaches 2 million pounds of high-purity uranium for Manhattan Project.[86] |
| 1947 | Ames Laboratory established by U.S. Atomic Energy Commission |
| 1950 | WOI-TV established as the first commercially operated television station owned by a university in the U.S. Station sold in 1994.[87][88] |
| 1954 | Cy becomes the Iowa State mascot |
| 1959 | Soviet leader Nikita Khrushchev visits Iowa State |
| 1959 | 10 kW, 150-ton nuclear teaching reactor is built. Reactor decommissioned and removed in 2000.[89] |
| 1959 | Renamed the Iowa State University of Science and Technology |
| 1959 | Iowa State's divisions become colleges: the College of Agriculture, College of Engineering, College of Home Economics, College of Sciences and Humanities, and College of Veterinary Medicine |
| 1962 | Enrollment reaches 10,000 students |
| 1966 | Enrollment reaches 15,000 students |
| 1968 | The College of Education is established |
| 1974 | The Maintenance Shop opens in the Memorial Union |
| 1979 | The College of Design is established |
| 1984 | The College of Business is established |
| 1988 | First VEISHEA Riot |

| 1992 | Second VEISHEA Riot |
| 1995 | Reiman Gardens opens[90] |
| 1997 | Working replica of Atanasoff-Berry Computer is unveiled, goes on nationwide tour[91] |
| 1999 | Central Campus is listed as a "medallion site" by the American Society of Landscape Architects |
| 2004 | Third VEISHEA Riot |
| 2005 | The College of Education and the College of Family and Consumer Sciences are combined to create the College of Human Sciences |
| 2006 | VEISHEA returns after being canceled for 2005; is deemed a huge success |
| 2007 | ISU's year long Sesquicentennial celebration is kicked off at VEISHEA 2007 with a 20,000-piece birthday cake[92] |
| 2008 | Sesquicentennial of Iowa State |
| 2009 | 25th Anniversary of the College of Business |

## See also

- C. Arden Pope
- CyRide
- Iowa State Center
- Reiman Gardens
- VEISHEA

## References


[1] http://www.foundation.iastate.edu/site/DocServer/ISUF_Financial_Statements_FY11_Final.pdf?docID=4221

[2] http://www.news.iastate.edu/news/2011/sep/fallenrollment

[3] http://www.news.iastate.edu/news/2011/sep/fallenrollment

[4] http://www.news.iastate.edu/news/2011/sep/fallenrollment

[5] http://www.iowastatedaily.com/news/article_aed0ad76-d9a0-11e0-99b9-001cc4c002e0.html

[6] http://www.iastate.edu/

[7] "Iowa State University" (http://classifications.carnegiefoundation.org/lookup_listings/view_institution.php?unit_id=153603& start_page=institution.php& clq={"ipug2005_ids":"","ipgrad2005_ids":"","enrprofile2005_ids":"","ugprfile2005_ids":"","sizeset2005_ids":"","basic2005_ids":"","eng2005_ids":"","se state","first_letter":"","level":"","control":"","accred":"","state":"","region":"","urbanicity":"","womens":"","hbcu":"","hsi":"","tribal":"","msi":"","landgra Carnegie Foundation for the Advancement of Teaching. . Retrieved December 20, 2010.

[8] History of Iowa State (http://www.lib.iastate.edu/spcl/exhibits/150/template/history.html). Iowa State University Website.

[9] History of Iowa State Time Line, 1858-1874. (http://www.lib.iastate.edu/spcl/exhibits/150/template/timeline-1858.html) Iowa State University Website.

[10] "Sesquicentennial Message from President" (http://www.public.iastate.edu/~isu150/president.shtml). Iowa State University. . Retrieved 8 September 2011.

[11] "Iowa State: 150 Points of Pride" (http://www.ag.iastate.edu/coa150/pop8_20.php). Iowa State University. . Retrieved 8 September 2011.

[12] History of Iowa State. (http://www.lib.iastate.edu/spcl/exhibits/150/template/time.html) Iowa State University Website.

[13] History of Iowa State Time Line, 1875-1899. (http://www.lib.iastate.edu/spcl/exhibits/150/template/timeline-1875.html) Iowa State University Website.

[14] History of Iowa State Time Line, 1900-1924. (http://www.lib.iastate.edu/spcl/exhibits/150/template/timeline-1900.html) Iowa State University Website.

[15] History of Iowa State Time Line, 1925-1949. (http://www.lib.iastate.edu/spcl/exhibits/150/template/timeline-1925.html) Iowa State University Website.

[16] State University Department of Computer Science Website. (http://www.cs.iastate.edu/jva/jva-archive.shtmllIowa)

[17] "ISU Naval ROTC - Unit History" (http://www.navy.iastate.edu/command/history.html). Ames, Iowa: Iowa State University. 2011. . Retrieved September 28, 2011.

[18]  History of Iowa State Time Line, 1950-1974. (http://www.lib.iastate.edu/spcl/exhibits/150/template/timeline-1950.html) Iowa State
      University Website.
[19]  Iowa State University History website (http://www.lib.iastate.edu/spcl/exhibits/150/template/history.html). Iowa State University
      Website.
[20]  Fall 2010 enrollment numbers (http://www.news.iastate.edu/news/2010/sep/2010enrollment)
[21]  "America's Best Colleges" (http://www.forbes.com/top-colleges/list/). Forbes. 2011. . Retrieved October 6, 2011.
[22]  "National Universities Rankings" (http://colleges.usnews.rankingsandreviews.com/best-colleges). *America's Best Colleges 2012*. U.S.
      News & World Report. September 13, 2011. . Retrieved September 25, 2011.
[23]  "The Washington Monthly National University Rankings" (http://www.washingtonmonthly.com/college_guide/rankings_2011/
      national_university_rank.php). *The Washington Monthly*. 2011. . Retrieved August 30, 2011.
[24]  "Academic Ranking of World Universities: Global" (http://www.shanghairanking.com/ARWU2011.html). Institute of Higher Education,
      Shanghai Jiao Tong University. 2011. . Retrieved August 30, 2011.
[25]  "QS World University Rankings" (http://www.topuniversities.com/university-rankings/world-university-rankings/2011). QS
      Quacquarelli Symonds Limited. 2011. . Retrieved September 30, 2011.
[26]  "Top 400 – The Times Higher Education World University Rankings 2011–2012" (http://www.timeshighereducation.co.uk/
      world-university-rankings/2011-2012/top-400.html). The Times Higher Education. 2011. . Retrieved October 6, 2011.
[27]  (http://www.carnegiefoundation.org/classifications/sub.asp?key=748&subkey=14302&start=782) Carnegie Classifications for Iowa
      State University
[28]  http://colleges.usnews.rankingsandreviews.com/best-colleges/ames-ia/iowa-state-university-1869
[29]  "Computer Engineering School Ranking" (http://www.uscollegeranking.org/engineering/
      top-100-graduate-computer-engineering-school-rankings-in-2011.html). 2011. . Retrieved 2011-03-22.
[30]  "About Greenlee School of Journalism and Communication" (http://www.jlmc.iastate.edu/about). . Retrieved 2011-08-31.
[31]  "Greenlee School Research Ranks in Top 30" (http://www.jlmc.iastate.edu/news/2010/summer/
      greenlee-school-research-ranks-top-30). 2010. . Retrieved 2011-06-03.
[32]  (http://www.ir.iastate.edu/FB11/PDF/FB11-020.pdf) ISU Fact Book 2010-2011. Iowa State University website. Retrieved 2011-3-24.
[33]  (http://www.lib.iastate.edu/libinfo/factsfigs.html) It's a Fact: Iowa State University. Iowa State University website. Retrieved
      2011-03-24.
[34]  (http://www.cs.iastate.edu/jva/jva-archive.shtml) About Iowa State University online
[35]  Vincent Atanasoff & the Birth of the Digital Computer" (http://www.cs.iastate.edu/jva/jva-archive.shtml|"John) Iowa State University
      Department of Computer Science Website.
[36]  History of Iowa State Time Line, 1975-2008. (http://www.lib.iastate.edu/spcl/exhibits/150/template/timeline-1975.html) Iowa State
      University Website.
[37]  (http://www.extension.iastate.edu/dubuque/info/history.htm) Dubuque County Extension History online
[38]  (http://www.iastate.edu/about/highlights/) About Iowa State University online
[39]  (http://www.iastate.edu/IaStater/1997/feb/landgrant.html) The Iowa Stater, February 1997
[40]  (http://www.ameslab.gov/overview/history.html) The Ames Laboratory website
[41]  It's a Fact: Iowa State University (http://www.iastate.edu/about/fact04/). Iowa State University website.
[42]  Gaines, Thomas (1991). *The Campus as a Work of Art*. New York: Praeger Publishers. pp. 155.
[43]  (http://www.lib.iastate.edu/arch/campanile/camphist.html) Iowa State University Library. "History of the Campanile". Retrieved
      2011-3-24.
[44]  (http://www.isualum.org/en/traditions/official_isu_ring/ring_symbolism.cfm) Iowa State University Alumni Association. "Ring
      Symbolism". Retrieved 2011-3-24.
[45]  (http://www.isualum.org/en/traditions/traditions_of_iowa_state/traditions_and_history/official_university_mace.cfm) Iowa State
      University Alumni Association. "Official University Mace". Retrieved 2011-3-24.
[46]  Swans (http://www.lib.iastate.edu/spcl/exhibits/VEISHEA/swans.htm) from the Iowa State Library's special exhibits section
[47]  http://www.news.iastate.edu/oldreleases/2003/mar/swans.shtml
[48]  (http://www.reimangardens.com/en/awards_and_publications/) About Iowa State University online
[49]  (http://www.museums.iastate.edu) Iowa State University Museums online. Retrieved 2011-03-10.
[50]  (http://www.museums.iastate.edu/Brunnier.htm) Iowa State University Museums Brunnier Art Museum online. Retrieved 2011-03-10.
[51]  (http://www.museums.iastate.edu/FarmHouse.htm) Iowa State University Museums Farm House Museum online. Retrieved
      2011-03-10.
[52]  (http://www.museums.iastate.edu/AOC.htm) Iowa State University Museums Art on Campus online. Retrieved 2011-03-10.
[53]  (http://www.museums.iastate.edu/CPAM.htm) Iowa State University Museums Christian Petersen Art Museum online. Retrieved
      2011-03-10.
[54]  (http://www.museums.iastate.edu/ASGMain.htm) Iowa State University Museums Anderson Sculpture Garden online. Retrieved
      2011-03-10.
[55]  "ISU promotes sustainability via an all-university compost facility" (http://www.public.iastate.edu/~nscentral/news/2009/may/
      compost.shtml). Iowa State University News. . Retrieved 2009-06-10.

[56] "New web site charts campus building energy use" (http://www.iastate.edu/Inside/2009/0417/energywatch.shtml). Iowa State University. . Retrieved 2009-06-10.

[57] "Live Green revolving loan fund" (http://www.livegreen.iastate.edu/loan/). Iowa State University News. . Retrieved 2009-06-10.

[58] http://www.greenreportcard.org/report-card-2011/schools/iowa-state-university

[59] The seven wonders of Iowa State (http://www.iowastatedaily.com/home/index.cfm?event=displayArticlePrinterFriendly& uStory_id=5e9b12e3-5c74-4628-b30d-9773a75eccc3). The Iowa State Daily.

[60] Government of the Student Body (http://www.gsb.iastate.edu/)

[61] (http://www.mu.iastate.edu/en/about_the_mu/architecture)"Architecture". Iowa State University website. Retrieved 2011-3-24.

[62] Memorial Union (Iowa State University) Retrieved 2011-3-24.

[63] "Greek Community Membership Statistics" (http://www.greek.iastate.edu/resources/Greek-Community-Statistics-Fall07.pdf) (PDF). Iowa State University Office of Greek Affairs. 2007-11-01. . Retrieved 2008-08-07.

[64] Alpha Chi Omega (2009). "Alpha Chi Omega" (http://www.alphachiomega.org/FIVE/ChapterProfile.aspx?contactID=100383). . Retrieved 2010-05-09.

[65] Pi Chapter. "ADPi" (http://www.alphadeltapi.org/pichapter/). Alpha Delta Pi Sorority. . Retrieved 2010-05-09.

[66] Rho Chapter (Alpha Gamma Delta) (2010). "Welcome to the Rho Chapter of Alpha Gamma Delta!" (http://www.isuagd.com/). Chapter Communications. . Retrieved 2010-05-09.

[67] Miller, W.J. (1961). "Greek Community Origins from Fraternities & Sororities at Iowa State" (http://www.greek.iastate.edu/history/ timeline.html). Iowa State University Office of Greek Affairs. . Retrieved 2008-08-07.

[68] The Scroll of Phi Delta Theta, Vol. XXXVII, (1912-1913) p 542, edited by Davis, T.

[69] Iowa State University Time Line, 1875-1899 (http://www.iastate.edu/~isu150/history/timeline-1875.html). Iowa State University website.

[70] Liwang, Roland (2007-06-01). "Most Dominant College Mascot on Earth: Cyclones win!" (http://www.sportsline.com/spin/story/ 10206409). Sportsline.com. . Retrieved 2008-06-06.

[71] "History of Iowa State: Time Line, 1875-1899. Iowa State University. 2007." (http://www.lib.iastate.edu/spcl/exhibits/150/template/ timeline-1875.html). .

[72] "Iowa State Media Guide-Records. 2008." (http://www.cyclones.com//pdf7/134422.pdf?SPSID=48396&SPID=4653& DB_OEM_ID=10700). .

[73] ""Iowa State Men's Basketball Media Guide" Iowa State University. 2008." (http://www.cyclones.com/ ViewArticle,dbml?DB_OEM_ID=10700&ATCLID=1608347&KEY=&DB_OEM_ID=10700&DB_LANG=&IN_SUBSCRIBE.html). .

[74] "Iowa State Bill Fennelly Bio. Retrieved June 2010." (http://www.cyclones.com/ViewArticle.dbml?SPSID=46668&SPID=4253& DB_OEM_ID..html). .

[75] "Cyclone Wrestlers Ready For NCAA's" (http://www.cyclones.com/ViewArticle.dbml?SPSID=46632&SPID=4248& DB_OEM_ID=10700&ATCLID=1413974). cyclones.com. 2008-03-18. . Retrieved 2008-06-09.

[76] VEISHEA History (http://www.veishea.iastate.edu/2006/mediakit/history.pdf) from the official 2006 media kit

[77] "Points of Pride - Alumni" (http://www.iastate.edu/about/alumni.shtml). Iowa State University. . Retrieved 2008-10-13.

[78] "History of Iowa State: Student Life" (http://www.iastate.edu/~isu150/history/student.html). Iowa State University. . Retrieved 2007-04-17.

[79] "Iowa State University Cemetery" (http://www.fpm.iastate.edu/cemetery/). Iowa State University. . Retrieved 2008-10-12.

[80] http://vetmed.iastate.edu/about/history

[81] "History of the College's Name" (http://www.ag.iastate.edu/news/namehistory.html). ISU College of Agriculture. . Retrieved 2008-10-12.

[82] http://www.veishea.iastate.edu/en/about_veishea/history_of_veishea/beginnings/

[83] "Figures from the History of Probability and Statistics" (http://www.economics.soton.ac.uk/staff/aldrich/Figures.htm). John Aldrich. . Retrieved 2008-10-14.

[84] "Inventors of the Modern Computer" (http://inventors.about.com/library/weekly/aa050898.htm). Mary Bellis. . Retrieved 2008-10-12.

[85] "John V. Atanasoff Dies at Age 91 Invented First Electronic Computer" (http://www.cs.iastate.edu/jva/jva-obit.shtml). Washington Post. . Retrieved 2008-10-12.

[86] "The 1940s — The Manhattan Project Years and After" (http://www.ameslab.gov/60thanniversary/1940.htm). Ames Laboratory. . Retrieved 2008-10-12.

[87] "Campus Journal; Vote to Sell TV Station Splits Iowans" (http://query.nytimes.com/gst/fullpage. html?res=9E0CE0D91E3FF931A15754C0A964958260). New York Times. 1992-07-22. . Retrieved 2008-10-12.

[88] "Iowans for WOI-TV, Inc." (http://www.lib.iastate.edu/spcl/manuscripts/MS584.html). Iowa State University Library. . Retrieved 2008-10-12.

[89] "Nuclear reactor removal underway at Iowa State U." (http://www.bookrags.com/highbeam/nuclear-reactor-removal-underway-at-hb/). University Wire. . Retrieved 2008-10-12.

[90] "History of Reiman Gardens" (http://www.reimangardens.com/index.cfm?nodeID=5725). Reiman Gardens. . Retrieved 2009-06-10.

[91] "Atanasoff-Berry Computer Replica Unveiled in Washington, D.C." (http://www.scl.ameslab.gov/ABC/Articles/DailyOct9-97.html). Iowa State Daily. . Retrieved 2008-10-16.

[92] "Lots of cake" (http://www.public.iastate.edu/~nscentral/news/2007/apr/0420.shtml). ISU News Service. . Retrieved 2009-02-12.

## External links

- Official website (http://www.iastate.edu/)

*This article incorporates text from an edition of the* New International Encyclopedia *that is in the public domain.*

# Troy_University

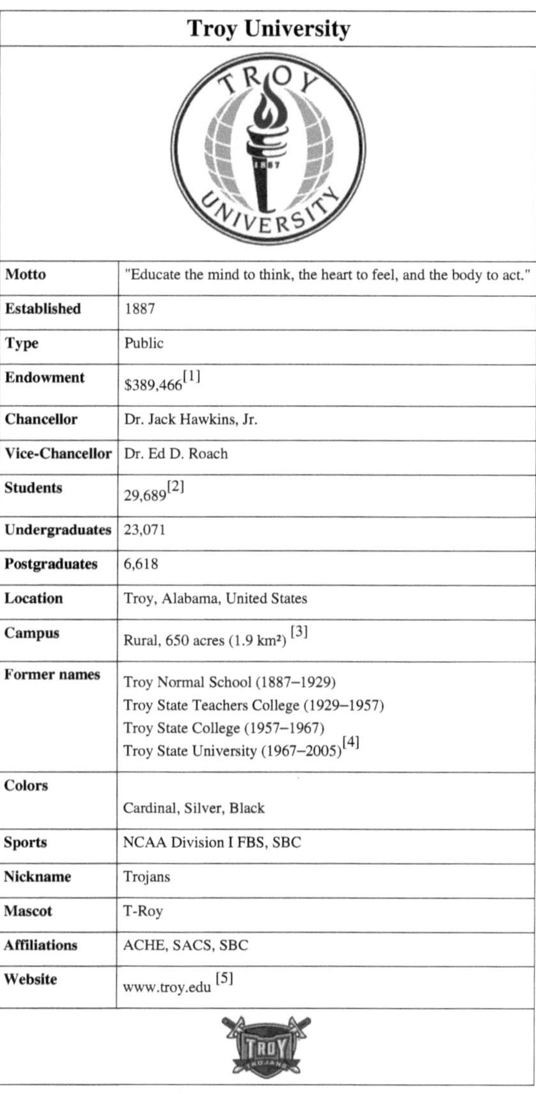

| Troy University | |
|---|---|
| Motto | "Educate the mind to think, the heart to feel, and the body to act." |
| Established | 1887 |
| Type | Public |
| Endowment | $389,466[1] |
| Chancellor | Dr. Jack Hawkins, Jr. |
| Vice-Chancellor | Dr. Ed D. Roach |
| Students | 29,689[2] |
| Undergraduates | 23,071 |
| Postgraduates | 6,618 |
| Location | Troy, Alabama, United States |
| Campus | Rural, 650 acres (1.9 km²) [3] |
| Former names | Troy Normal School (1887–1929)<br>Troy State Teachers College (1929–1957)<br>Troy State College (1957–1967)<br>Troy State University (1967–2005)[4] |
| Colors | Cardinal, Silver, Black |
| Sports | NCAA Division I FBS, SBC |
| Nickname | Trojans |
| Mascot | T-Roy |
| Affiliations | ACHE, SACS, SBC |
| Website | www.troy.edu [5] |

**Troy University** is a public university that is located in Troy, Alabama, United States. It was originally founded in 1887 as Troy Normal School. Its main campus enrollment is 7,194 students. The total enrollment of all Troy University campuses is 29,689. Today, Troy University offers bachelor's, master's, and doctoral degrees

# History

Troy Normal School

Troy University is a public university with its main campus located in Troy, Alabama. It was founded as a normal school in 1887 with a mission to educate and train new teachers. The school has since evolved into a state university, located in four sites across the Alabama: Troy, Montgomery, Phenix City and Dothan. The university also has sites located throughout the United States and in international locations. Troy is known for both in class and online academic programs and service to traditional, nontraditional and military students. The main campus enrollment as of the fall of 2010 is 7,194 students.[4] The campus itself consists of 36 major buildings on 650 acres (1.9 km²) plus the adjacent Troy University Arboretum. [edit]

## Name change

On April 16, 2004, the Board of Trustees voted to change the name of the institution from **Troy State University** to **Troy University**. The transition to the new name was completed in August 2005. The name change was the fifth in the school's history. When created by the Alabama Legislature on February 26, 1887, it was officially named the **Troy State Normal School**. The school was located in downtown Troy until moving to the present location on University Avenue in 1930. In 1929, the name was changed to **Troy State Teachers College** and it subsequently conferred its first baccalaureate degree in 1931. In 1957, the legislature voted both to change the name to **Troy State College** and to allow it to begin a master's degree program. The name was changed once again in 1967 to **Troy State University**. In 2010, the University began offering its first doctoral degree, the Doctor of Nursing Practice.

# Academics

## Structure

Troy University cumulatively offers 46 bachelor's degree programs, 22 master's degree programs, and one doctoral program with an overall selective acceptance rate of 68%.

The University is composed of:

- Five Colleges
- The Graduate School
- The Division of General Studies

Hawkins Hall on the Troy University campus.

## Rankings

| University rankings (overall) | |
| --- | --- |
| **National** | |
| *Forbes*[6] | 327 |
| **Regional** | |
| *U.S. News & World Report*[7] | 63 |
| **Master's University class** | |
| *Washington Monthly*[8] | 424 |

Troy University has acquired different institutional rankings from various sources:

- Forbes Magazine ranks Troy among the top tier of institutions. Forbes' overall ranking centers on the value of the degree obtained by a university's students and measures, in part, the marketplace success of a school's graduate.

- The US News & World Report has Troy University ranked as the #63 *Regional University* in the southeastern United States, as well as being ranked the #28 *Top Public School* in the nation.

- In the US News & World Report's *Best Grad School* national rankings, Troy is listed as having the #83 Rehabilitation Counseling program, along with the #442 ranked Nursing program.

- For the seventh year in a row, the Princeton Review named Troy to its "Best in the Southeast" list due to excellence in academic programs and institutional data collected from the university.[9]

- In 2008, Troy University was ranked as the 25th best university in the United States for international students by the Institute of International Education.

## Campus

Troy University's campus is located near downtown Troy. The campus sits along rolling hills with many old oak trees present along the streets and throughout campus. A very small lake named *Lake Lagoona* sits in front of university's Sorrell Chapel.

The university closed the **Trojan Oaks Golf Course** in order to build a new basketball arena, but remnants of the course are still present and are utilized freely for practice.

Troy University buildings in downtown Montgomery

Main Campus at Troy University

# Student life

Part of the Trojan Village student housing complex on the Troy University campus.

Bibb Graves from University Avenue on the Troy University campus.

| | |
|---|---|
| **Alumni Hall** | Men |
| **Clements Hall** | Coed |
| **Cowart Hall** | Women |
| **Gardner Hall** | Women |
| **Hamil Hall** | Women |
| **Hillcrest Hall** | Men |
| **Honors Cottage** | Coed |
| **Pace Hall** | Coed |
| **Paden Hall** | Women |
| **Shackelford Hall** | Coed |
| **Trojan Village** | Coed |
| **University Apts** | Coed |

Students who live on campus at Troy have a variety of housing options. There are 12 halls and each hall offers different amenities and communities.

## Dining

The **Trojan Center** and the **Trojan Dining** buildings offer a wide-variety just of dining options for students.

The Trojan Dining hall on the Troy University campus.

**Trojan Center:**

- Chick-fil-A
- Sub Connection
- A&W Restaurant
- Mein Bowl
- Einstein Bros. Bagels
- Herb's Place
- Starbucks

**Trojan Dining:**

- Moe's Midwest Grill
- Boar's Head Deli
- Flying Star
- Bella Trattoria
- Mongolian Grill
- Wildest Mushroom
- Basic Kneads
- Magellan's Restaurant.

## Recreation

There are many recreational activities available on campus. The **Trojan Fitness Center** offers fitness machines, free weights, and cardiovascular machines. **Trojan Games** is a recreation room has two billiard tables, one table tennis tables, mounted televisions, and a video game system. The **Natatorium** houses an eight-lane 25 yard Olympic style pool. The **Recreation Center Gym** has two basketball courts, cardio room, a dance room, and a large outside pool. **Wright Hall Gym**, located adjacent from the Natatorium, offers a basketball court, two volleyball courts, and four badminton courts. The **Intramural Fields** consist of four flag football fields, two softball fields, and one soccer field.

There are plans in place to renovate **Sartain Hall**, which currently in its last season as home to the men's and women's varsity basketball teams. Currently, it is expected to be turned into a large recreational facility.

## Transportation

The Troy University Transportation Service offers students a free shuttling service across campus at various stops across campus.

## Greek Life

There are 21 traditional Greek organizations on Troy's campus.

| NIC Fraternities[10] | Alpha Tau Omega (ATΩ) | Delta Kappa Epsilon (ΔKE) | Delta Chi (ΔX) | Lambda Chi Alpha (ΛXA) | Tau Kappa Epsilon (TKE) | FarmHouse (FarmHouse) | Pi Kappa Phi (ΠKΦ) | Sigma Chi (ΣX) |
|---|---|---|---|---|---|---|---|---|
| NPC Sororities[11] | Alpha Gamma Delta (AΓΔ) | Alpha Delta Pi (AΔΠ) | Chi Omega (XΩ) | Phi Mu (ΦM) | Kappa Delta (KΔ) | | | |
| NPHC Greek Life[12] | Alpha Kappa Alpha (AKA) | Zeta Phi Beta (ZΦB) | Delta Sigma Theta (ΔΣθ) | Phi Beta Sigma (ΦBΣ) | Sigma Gamma Rho (ΣΓP) | Omega Psi Phi (ΩΨΦ) | Kappa Alpha Psi (KAΨ) | Alpha Phi Alpha (AΦA) |

## Religious and other organizations

The university is home to numerous religious campus organizations such as: The Campus Awakening, Chi Alpha, Wesley Foundation, Baptist Campus Ministries, Pentecostal Campus Ministries, The Newman Center (Roman Catholic), The Christian Student Center (Church of Christ), and The Troy Secular Association. Some of these religious organizations have stand-alone physical facilities on the Troy campus.

There are five Greek organizations that function under the supervision of the John M. Long School of Music.

Sorrell Chapel.

**Music Organizations at TROY:**

Phi Mu Alpha

Kappa Kappa Psi

Sigma Alpha Iota

Tau Beta Sigma

Phi Boota Roota

NAACP

## Student Media

The school newspaper, the Tropolitan (commonly referred to as "The Trop"), is located on the bottom floor of Wallace Hall. It is a weekly publication, written and produced entirely by students. The Palladium, is located in adjacent offices in the same building. Also located in Wallace Hall is Troy University Television, also referred to as Troy TrojanVision. Troy University Television broadcasts two live entirely student produced newscasts twice daily.

## The "Sound of the South" marching band

Music is an integral part of Troy University. The university boasts 29 faculty in the School of Music, over 200 undergraduate music majors, and fields a variety of music ensembles including a Symphonic Band, two Concert Bands, two Jazz Bands, a Trumpet Ensemble, Pep Band, Brass Quintet, and more. The school of music also hosts a Brass Symposium every spring semester.

The Sound of the South plays halftime shows at all Troy home football games and many of the away games. The band is noted for traveling as much as the football team, some recent trips were University of Nebraska, University of Miami, University of Arkansas, Mississippi State University, University of Florida, University of Georgia, and Florida State University. The Sound has been featured at numerous bowl games, including the Peach Bowl, the Senior Bowl, the Blue-Gray Football Classic, the R+L Carriers New Orleans Bowl, and has also been featured in halftime performances for the Atlanta Falcons, Miami Dolphins, New Orleans Saints and Tampa Bay Buccaneers. Over the past two years, the band has performed for over 450,000 fans. The band has recorded for the Warner Brothers Marching Band Promotional Compact Disc since 1998, which is distributed to over 38,000 bands. The Sound of the South performs regularly at Veterans Memorial Stadium in Troy, Alabama.

## Athletics

Troy State Normal School began its sports program in 1909 when it fielded its first football team. Through the early years Troy's athletics nicknames were not official and varied by the sport and the coach. Eventually teams all began to use the name "Troy State Teachers," but when the athletic teams moved into NAIA competition the nickname was then was changed to the "Red Wave". In the early 1970s the student body voted to change the name to Trojans after many felt that Red Wave was too similar to the University of Alabama's nickname, the Crimson Tide. Prior to becoming a member of NCAA Division One athletics in 1993, Troy University was a member of the Gulf South Conference of the NCAA Division II ranks. Troy's primary rivals were Jacksonville State University, Livingston University (now the University of West Alabama), and the University of North Alabama.

Troy Football Team 2009

### Football

TROY Athletics

Troy University began playing football in 1909. The program has won three national championships, the NAIA national football championship in 1968 and the NCAA Division II national football championship in 1984 and 1987. Troy transitioned to the NCAA's Division I-A in 2002, became a football only member of the Sun Belt Conference in 2004, and joined the conference for all other sports in 2005. In 2001, Troy defeated Mississippi State at Scott Field in Starkville, Mississippi, by a score of 21–9 which was the Trojans' first victory over a BCS level program. In 2004, the Trojans defeated a ranked BCS program for the first time ever, defeating #17 Missouri 24-14 at home on ESPN2. The Trojan football team made its first bowl game appearance in the Silicon Valley Football Classic on December 30, 2004, but lost to Northern Illinois, 34–21. In 2006, Troy won the Sun Belt Conference for the first time after defeating Middle Tennessee State toward the end of the 2006 season. Troy represented the Sun Belt Conference in the 2006 New Orleans Bowl as the conference champion for the first time where the Trojans defeated the Rice Owls of Conference USA by a score of 41–17. Troy has most recently participated in the 2010 New Orleans Bowl where the Trojans routed Ohio by a score of 48–21. Troy football head coach Larry Blakeney is entering his 20th season as head coach. He has led the program to three Southland Football Conference titles and five Sun Belt Conference titles, as well as guided the Trojans to seven FCS playoff appearances and five FBS bowl games. Blakeney boasts an overall record of 161–80–1 as head coach at Troy. Blakeney is the winningest coach in the Troy University history and he is the 4th winningest collegiate coach all-time in the state of Alabama, only behind greats Paul "Bear" Bryant, Cleve Abbott, and Ralph "Shug" Jordan. Blakeney is one of two coaches in college football history to be the head coach of a football program during its transition from Division II to I-A (the other being UCF's Gene McDowell).

Veterans Memorial Stadium, home of Trojan Football

## Basketball

The Troy University men's basketball team is under the direction of head coach Don Maestri. Head coach Don Maestri is the winningest coach in Troy University history and he has won numerous conference coach-of-the-year awards during his tenure at Troy University. The program has won 11 conference championships in basketball, with six of them coming in the Division I era. On January 12, 1992, Troy defeated DeVry University of Atlanta by the score of 258–141, in what is the highest scoring game in college basketball history. The Trojans competed in the 2003 NCAA Tournament in Nashville against Xavier University after winning the Atlantic Sun Conference title. In 2004, Troy was an NIT participant in a match-up against Niagara University. In 2009, the Trojans finished 3rd place in the Sun Belt Conference and competed in the CBI against College of Charleston. After winning the Sun Belt regular season title in 2010, the Trojans would be invited to play in the NIT once again against Ole Miss. In 2008, Don Maestri was inducted into the Wiregrass Sports Hall of Fame at a ceremony in Dothan, Alabama. The Troy University women's basketball team is currently under the direction of head coach Michael Murphy. The women's basketball program last competed in the NCAA tournament in 1997 against the University of Virginia under the direction of head coach, Jerry Hester.

| Troy University Fight Song |
| --- |
| Here's to the school we love |
| We are Trojans, one and all |
| We will always cheer for victory, |
| And you'll never let us fall |
| GO! GO! GO! |
| Cheers to T-R-O-Y |
| We are with you all the way |
| So get out there team |
| And FIGHT FIGHT FIGHT |
| And win today |

In 2008, Don Maestri was inducted into the Wiregrass Sports Hall of Fame at a ceremony in Dothan, Alabama. The Troy University women's basketball team is currently under the direction of head coach Michael Murphy. The women's basketball program last competed in the NCAA tournament in 1997 against the University of Virginia under the direction of head coach, Jerry Hester.

## Baseball

The Troy University baseball team won two Division II national championships in 1986 and 1987 under the leadership of coach Chase Riddle. One of Troy's biggest victories in baseball came in April 1998 when the Trojans knocked off the #3 nationally ranked University of Alabama Crimson Tide by a score of 8–4 at Riddle-Pace Field on the Troy campus. Under the direction of current head coach Bobby Pierce, the Trojan baseball program has competed in the NCAA Baseball Tournament in 2006 and 2007. Troy also competed in the 1995 and 1997 NCAA Division One tournament under head coach John Mayotte. In 1999, the program broke the NCAA Division 1 record for most hits in an inning, belting 14 hits in a 34–4 rout of Stetson.

## Rodeo

The program's governing body is the National Intercollegiate Rodeo Association. The rodeo program's home facility is the Pike County Cattlemen's Arena in Troy where it hosts a three-day rodeo each October that features college rodeo programs from throughout the southern region of the United States. Troy University calf roper Ben Mayworth won the 2007 national title in Casper, Wyoming, at the National Finals Collegiate Rodeo.

# Campus/Academic Features

The international student living/learning center, Pace Hall.

### Hall of Fame of Distinguished Band Conductors

Main article: National Band Association Hall of Fame of Distinguished Band Conductors The Hall of Fame of Distinguished Band Conductors was established on the campus of what was then known as Troy State University in Troy, Alabama by the National Band Association in 1979. The Hall of Fame contains the picture and biographies of band directors who have distinguished themselves in some way or who have made significant contributions to the field of band directing, conducting, or leadership.

### Confucius Institute

The Confucius Institute is a non-profit public institute that aims at promoting Chinese language and culture and supporting local Chinese teaching internationally through affiliated Confucius Institutes. Its headquarters is in Beijing and is under the Office of Chinese Language Council International or Hanban. The first Confucius Institute in the state of Alabama opened at Troy University in the fall of 2008.

The Adams Administration Building, located on the Quad.

### The Manuel H. Johnson Center for Political Economy

Troy University's Manuel H. Johnson Center for Political Economy was formed in September as the result of a $3.6 million gift from Troy alumnus Dr. Manuel H. "Manley" Johnson, BB&T bank and the Charles G. Koch Charitable Foundation. The Center's mission is the advancement of free market economic ideas and its research and teaching efforts explore the idea that economic freedom improves the quality of life for citizens. The new Center will be part of the University's Sorrell College of Business and will be housed inside Bibb Graves Hall. Dr. Scott Beaulier serves as the center's executive director.

## Notable alumni

Mike Rivera, 1997   William G. Gregory, 1984   Bobby Bright   Manuel H. Johnson, 1973   Kevin R. Kregel, 1988   DeMarcus Ware, 2005

Osi Umenyiora, 2003   Lawrence Tynes, 2001   James A. Roy, 2000   Leodis McKelvin, 2008   Husky Harris   Marshall B. Webb, 1994

## References


[1] As of June 30, 2010. "Best Colleges 2012: Troy University" (http://colleges.usnews.rankingsandreviews.com/best-colleges/troy-university-1047). U.S. News and World Report. . Retrieved January 6, 2012.

[2] (http://www.troy.edu/irpe/archives/2007-2008/2008-factbook/Fact-Book-2008.pdf)

[3] http://events.wsfa.com/Troy_University/v175228212.html

[4] Songe, Alice. *American Universities and Colleges: A Dictionary of Name Changes.* Scarecrow Press (Metuchen, NJ: 1978), p. 213

[5] http://www.troy.edu

[6] "America's Best Colleges" (http://www.forbes.com/top-colleges/list/). Forbes. 2011. . Retrieved October 6, 2011.

[7] "Regional Universities Rankings" (http://colleges.usnews.rankingsandreviews.com/best-colleges/rankings/regional-universities). *America's Best Colleges 2012.* U.S. News & World Report. September 13, 2011. . Retrieved September 25, 2011.

[8] "The Washington Monthly Master's University Rankings" (http://www.washingtonmonthly.com/college_guide/rankings_2011/masters_universities_rank.php). *The Washington Monthly.* 2011. . Retrieved October 6, 2011.

[9] http://www.troy.edu/news/archives/2011/august/08092011_best_southeast.html

[10] http://troy.troy.edu/ifc/fraternities.html Interfraternity Council

[11] http://troy.troy.edu/organizations/sororitylife/index.html College Panhellenic Council

[12] http://troy.troy.edu/nphc/thedivinenine.html National Pan-Hellenic Council

## External links

- Troy University Main Campus (http://www.troy.edu/)
- Troy University Athletics (http://www.troytrojans.com/)
- Troy University eCampus (http://www.troy.edu/ecampus/)
- Troy University Dothan Campus (http://dothan.troy.edu/)
- Troy University – Montgomery Campus (http://montgomery.troy.edu/)
- Troy University – Phenix City Campus (http://phenix.troy.edu/)
- Troy University Official Newspaper of Troy University (http://trop.troy.edu/)
- Chinese Station at Troy University (http://bbs.imence.com/)

# Article Sources and Contributors

**American_football**  *Source*: http://en.wikipedia.org/w/index.php?title=American_football  *Contributors*: (aeropagitica), 1-555-confide, 128.32.172.xxx, 16@r, 1995hoo, 1sttomars, 2, 207.44.114.xxx, 24Seven, 2D, 3sn, 77hg66, A.C. Norman, A18919, A2Kafir, A314268, A3RO, ABF, AStudent, AaronY, Aaronp808, Abductive, Accurizer, AceKingQueenJack, Acroterion, Adambro, Addaone, Adderz, Addshore, Aeric67, Aerion, Afasmit, AgentPeppermint, Agentsirus, Ags412, Ahoerstemeier, Aillema, Aitias, Albanaco, Ale jrb, Alecsdaniel, Alex Lin, Alex.muller, Alexandria, Alexius08, AlexiusHoratius, Alexmetal07, Alias Flood, Aliengruvgod, Aliza250, Alki, AllPurposeGamer, Allstarecho, Almega123, Amalthea, Amavel, Amelansn, Amerikan, AmishThrasher, Ampersand692, Anatidae, Andi15ro, Andonic, Andre Engels, Andres, Andrew Levine, Andrewmarcum, Andries, Andros 1337, Andy56Iwoh, Angryapathy, Anilingus, Anomaly2002, Antandrus, Antiedman, Antimatt, Antonio Lopez, AntonioMartin, Anubis640, Apatterno, Apparition11, Arcadian, Arch dude, Arch26, Archanamiya, Arcimpulse, AreYouTheGateKeeper, ArglebargleIV, Arkangel lucifer, Armyrifle9, AscendedAnathema, Aspensti, Astral, Atlant, Atticust, Avant Guard, Avenged Eightfold, Avenue, Avicennasis, Awesomejc13, Awhite3, AxelBoldt, Ayrton Prost, Az1568, B, B-1yeti, B4hand, Badass llama, Bakabaka, BalthCat, Banes, Baqu11, Bardeousis, Barek, Bark, Bart133, Bartolofartolo, Bazonka, BazookaJoe, Bbpen, Bcisys, Bdb484, Bdesham, Bdizzy19, Beach blvd, BeastRHIT, Beaver, Beckford14, Beeanny, Beeblebrox, Ben Arnold, Ben-Zin, Benanhalt, Benbread, Bencherlite, Bender235, Benhagerty, Bennybp, Bensimeon, Betaeleven, Betterusername, Bettia, Bevo74, Bevo873, Beyond silence, Bfigura's puppy, Bgwwlm, Bhrdkor123, BiggKwell, Bigrich, Bihco, BilCat, BillNiTheScienceGuy, Birczanin, Bismarck, Bkell, Bkporter12, Blaque777, Blast22, Blavergn, Bletch, Blind justice, Blitzburg701, Bloody whore, BlueGoose, Bluelion, Bluemask, Bmstephany, Bob K31416, Bob rulz, Bob2688, Bobblewik, Bobby131313, Bobbyd1234, Bobbymozza, Bobo192, Bongwarrior, Bookermorgan, Booyabazooka, Boris Barowski, Bornintheguz, Bostonian Mike, Boznia, BradBeattie, Brainyiscool, Brand35, Brandonrush, Breno, Brholden, Brianyoumans, BrokenSphere, BruceDLimber, Brusegadi, Brutaldeluxe, BruteTAH, Brycejd1, Bs4173, Bschorr, Buchanan-Hermit, Buckswin82, Bughouse26, BuickCenturyDriver, Bunghole22, Burner0718, Burntsauce, Buzz-tardis, Bvs0001, C.Fred, CBrady15, CJSalerno, CO, CPAScott, CPacker, CStyle, CWY2190, CWii, CalJW, Calabraxthis, Calistemon, Caltas, CambridgeBayWeather, Can't sleep, clown will eat me, CanadianLinuxUser, Canterbury Tail, Captmjc, Carbonite, CardinalDan, Cardsplayer4life, Carlroller, Carrp, CarstenBN, Cconnett, Cdogberg, Cenarium, Central Data Bank, Centrx, Cesarcossio, Chainclaw, Chandler, Changtaiyeh, CharlieZeb, CharlotteWebb, Charlotteriversdale, ChazBeckett, Chemtiger05, Chensiyuan, Chillum, ChiragPatnaik, Chitrapa, Choster, Chowbok, Chris the speller, ChrisfromHouston, Chrislk02, Christensen28, Christian List, Christopher Parham, Chuck Sirloin, Civil Engineer III, Cje, Ckatz, Claidheamohmor, Cleared as filed, Click23, Clintheacock66, ClockworkLunch, Closedmouth, Cman736, Cocoaguy, ColinJF, Colorado avalanche, CommonsDelinker, Congerwiki, Connah0047, Conniexo, Conversion script, Coolcaesar, Corvus cornix, Corvus13, Costlab, Cowboy357, Crabula, Crash Underride, CrazyChemGuy, Crazycomputers, Crebbin, Cremepuff222, Crissov, Cs-wolves, Cshay, Ctbolt, Cuchullain, Curps, CyberSach, Cyrius, D, D-Rock, D. Webb, D6, DAOsh, DCHoya, DJ Clayworth, DMacks, DPMulligan, Daa89563, DabMachine, Dabigtrain, Daddy Kindsoul, Dale Arnett, DancingPenguin, DandyDan2007, DangApricot, Daniel, Daniel11, Danthemang531, DarkFireTaker, Darkspots, Darth Panda, Dave Foley, Dave101, DaveSDCali, Daveb, David of Earth, DavidLevinson, DavidSteinle, Davidmayberry, Dawn Bard, Dawson0782, Dawynn, Db099221, Dbarnes99, Dbfirs, Dbpride13, Dcatunlucky, DeadEyeArrow, Debigboy, Debresser, Decrypt3, Delgado in Color, Delirium, Delldot, Denelson83, Denis Moura dos Santos, DerHexer, Derek Ross, Dgrant, Dhmachine31, Dhp1080, Diceslot88, DickieRay, Diggsjj, Digital20, Discospinster, Diving2010, Djdickmutt, Djln, Dknights411, Dlayersjr, Dmr2, Dnvrfantj, DocWatson42, Dodiad, Domitius, Doopey, Doops, Doprendek, DougsTech, Doyley, Dpiranha, Dposse, Dragonfury, Dreaded Walrus, Dreamafter, Drsayis2, Dryazan, Dsmith113093, Dthomsen8, Duck16, Dukeofomnium, Dumbo3k, Duncharris, Durin, DurotarLord, Dwheeler, Dysepsion, Dysprosia, ESkog, Eagle4000, EarlyBird, East718, Easybake92, Ed Poor, Ed g2s, Edward Morgan Blake, Edward321, Edwardlalone, Edwinstearns, Eeekster, Egern, Ekotekk, ElKevbo, ElTyrant, Eleland, Eliasm13, Elipongo, Eliz81, Ellsworth, Emilyxxapple, EmirA, Emmarosman, Emurphy42, EngineerScotty, Enigmaman, Enviroboy, Eob, Epbr123, Epolk, Equiquinos, Eravau, Eric080, Erikdw, Erikshadow, Esanchez7587, Esperant, Estarriol, Esv216, Eternal Pink, Etienne.navarro, Ettrig, Euchiasmus, Euku, Evadb, EvelinaB, EveryDayJoe45, Everyking, Everything counts, Evil Monkey, Evilbob2, Evilweevil, Excirial, Excitinginterception7, Exploding Boy, Extransit, Ezn, FC Ownage99, FF2010, FRHS, Fabrictramp, Facts707, Faithlessthewonderboy, Falcofire, FalseAxiom, Familydz, Fang Aili, Fanman11, Fantumphool, Farosdaughter, Farside268, Fasttimes68, Faulcon DeLacy, Feedmecereal, Ferkelparade, Fibonacci, Finalius, Fingers-of-Pyrex, Finngall, Fishal, Fitch, Flakinho, Flamerule, Flapeyre, Flash19901, Flipsterz4life, Flowerpotman, Flubeca, Flyguy33, Flynntastico, Foolip, Football sucks, Football101, FootballGod, Footballplayer40, Fordan, Forteblast, FrWaters, Frankbar23, Freakofnurture, Fred Bauder, Freddiem, Frehley, Fremsley, FreplySpang, Friday, Fritsky, Fudoreaper, Fui in terra aliena, FullMetal Falcon, Funnyhat, FurmanUSC, Furrykef, FutureNJGov, Fuzheado, Fylc, Fyre2387, GABaker, Gadfium, Gaius Cornelius, Ganon2, Gardgate, Garfield226, Gary King, Garyzx, GeneralDuke, Geni, Geniac, Gentgeen, Geoffrey Gibson, Georgia guy, Gerry D, Gettingtoit, Gilliam, Giraffedata, Girdi, Gjd001, Gladboy101, Gmatsuda, Gnevin, Gogo Dodo, Golbez, Golfman5000, Gonzo fan2007, Goodluckme, GordyB, Gowikipedia, Gpietsch, GraemeL, GraemeLeggett, Grafikm fr, Grant Gussie, Grant65, Grapes911, Greecepwns, Green meklar, Gregbard, GregorB, Grondemar, Grstain, Gsmalls, Gtg204y, Gummi key, Gunsnroses15, Gurch, Gurchzilla, Gwernol, H2O, Hadal, Haham hanuka, Hakufu Sonsaku, Haldraper, Haleyisawesome, HappyJake, Hardworker111, Hateless, Havoc1310, Hcheney, Hdante, Hdt83, Headcoachhall, Hellno2, Henry W. Schmitt, HenryLarsen, Hephaestos, Herdrick, Hersfold, Heyes moo, HiDrNick, Hig Hertenfleurst, Hlynz, Hmains, Hment, Hmoul, Hog70, Hst43, Hsukicker, Hughs, Husond, Hux, Hydrogen Iodide, ICloudz1, II MusLiM HyBRiD II, IJK Principle, IRP, IRelayer, Iamdadawg54454, IanManka, Iankap99, Icairns, Iceberg3k, Ilikeeatingwaffles, Imaboy1235, Imkoolz, Indon, Informedopinion, Insanephantom, Instinct, Inter16, Introw, Ioeth, Iridescent, IrisKawling, Irish Souffle, Irishguy, Ironbearg, Isaiahcambron, Isozvaillancourt, Ivansevil, J-boogie, J-stan, J. Ngtson, J.delanoy, JAF1970, JCO312, JDspeeder1, JFreeman, JHunterJ, JIP, JMyrleFuller, JNW, JPD, JPMJPMJPMJPM, JS.Farrar, JackLumber, Jackfork, Jacob jordan, Jacob.jose, Jacoplane, Jacqui M, Jag1957, Jahiegel, Jamacfarlane, James26, James3167, JamesMLane, JamesTeterenko, Jamesfett, Jamesofur, Janothird, Jaranda, Jarbear800, Jaronsampson, JasonJNoble, Jauerback, JaySmo, JaySoldier3, Jayron32, Jayshao, Jbeau18, Jcam, Jcgarcow, Jclemens, Jcmenal, JdwNYC, Jebba, Jeff3000, Jeff79, Jennavecia, Jerzy, Jestersinthemoon, JesusChristismysavior, Jfdwolff, Jfiling, Jfitzg, Jgera5, Jhamby, Jhfireboy, Jimothytailor, Jiy, Jjron, Jnc, Jnelson09, Jnestorius, JoaoRicardo, Jock Boy, JockSoFine, Joehaer, Joeiii63, John Anderson, JohnWittle, Johnmercury007, Johnsonkurtis, Johntex, Joke137, Jonemerson, JorgeGG, Josh Myers-Jogn, Joshdboz, Joshparmour, Joshua Scott, Joshuapaquin, Josmcho101, Josquius, Jossi, Jpgordon, Jrkarp, Jrt989, Jsc1973, Jsnell, Jsylvest, Jtmoon, Juliancolton, Jump01, Jusdafax, Justinbaker, Justinhaynes, Jutta, Juzeris, Jweiss11, Jwy, Jxg, KCinDC, Kahuna936, Kainaw, Kal5108, Kaldari, Karmafist, Katalaveno, Katieh5584, Ke4roh, Ke5crz, Keegan, Keenan Pepper, Keilana, KelleyCook, KelvSYC, Kenkool16, KennethUrban, Kenny7070, KenyaDaMotherland, Kerowren, Ketiltrout, Kevin W., Khoikhoi, King of Hearts, Kinglehr, Kingpin13, Kingturtle, Kirjtc2, Kiske, KitHutch, Kitiara99, KittenKrazy, KiwiJeff, Kmweber, Koavf, Kodachromie, Kolakowski, Kornfan71, Korny O'Near, KramarDanIkabu, Krellis, Kristof vt, Krushdiva, Kryzty, Kshahn, Kubigula, Kukini, Kungfuadam, Kuru, Kuyabribri, Kwekubo, Kyunghoonie, LMA, Lacrimosus, Lancia66, LaruaWA11, Latics, Laukster, Lawrence Cohen, LeaveSleaves, LeeZ, Lefty, Legendlee, Lenoxus, Leonard G., LeonardoRob0t, Leonidasthespartan, Lesserm, Levineps, Liamlatics, Lightmouse, Ligulem, LilHelpa, Lilac Soul, Liquidvelvet, Little Mountain 5, LizardWizard, Lizmarh, Logan, Lomn, Londo06, LonelyMarble, Looper5920, Lotje, Lowellian, Lrc-101, Lsommerer, Luckett94, Luna Santin, Lunarbunny, Lupo, Lyonsonly, MD1937, MER-C, MK8, MKTFC, MLA, MMAfan2007, MZMcBride, MacRusgail, Mack2, Maduskis, Maid4TV, Mailer diablo, Majorclanger, Malinaccier, Malinaccier Public, Manburger 486, Marc Kupper, Marc-Olivier Pagé, MarcoAurelio, Marcus1060, Mareino, Mark Taylor, Markb, Masonpatriot, Massmissile, MateoCorazon, Mato, Matt Gabriel, Matt McDonald, Matt Yeager, Matt smoke 2, Mattbr, Matthead, Matthearn, Mattworld, Mav, Maxim, Maximus Rex, Maxw11, Mayn Payge, Mayumashu, Mccardle28, Mcgagne0, Mebaeta, Mechjerk, Meegs, Mendaliv, Menthaxpiperita, Mets4117, Mgaved, Michael Snow, Michael.gierasimiuk, Michael07lu, Mike dantonio14, Mike2vil, Miked9210, Milkncookie, Mirer, MisfitToys, Mister Alcohol, MisterCharlie, MisterHand, MisterStabledDS, Mithotyn, Mitul0520, Mllambring, Mmdolbow, Moanzhu, Modify, Modulatum, Moncrief, MonkeyMumford, Moondyne, Moonslight, Mooooooonydoggger, Moralis, Moreschi, Mormegil, MorrisS, Morrowkp1, Moverton, Mr Stephen, MrJ, MrMurph101, MrVibrating, Mrbrownshoes, Mrcandy, Mrmiscellanious, Mrschimpf, Mrsmith123, MsDivagin, Mtorpey, Muenda, Muhandes, Mwalcoff, N00b 0wange, NBcoolster, NCurse, NHRHS2010, NJA, NYScholar, Nabablucknow, Nabla, Nagy, Nat682, NawlinWiki, Ndk1134, Nehrams2020, Nengli02, Nepenthes, New World Man, NewEnglandYankee, Next362, Nick81, Nickknx865, Nickshanks, Nicksname, Nihiltres, Nikai, Nil Einne, Nishkid64, NiteowlneiIs, Nivix, Nlu, NorsemanII, Novacatz, Nsaa, NuclearWarfare, Nurg, Nutritionfan, Nuttycoconut, Obmatjd, Octane, Od Mishehu, Oda Mari, Oddball600, Oguz1, Ohnoitsjamie, Ok145897, Okiefromokla, OldRightist, Oldelpaso, Omicronpersei8, Onebyone, Oore, Optimist on the run, Opus33, Oreo Priest, Oroso, Ortolan88, Ositadinma, Ospalh, Owen, Oxymoron83, Ozgod, PCock, PDXblazers, PGPirate, PJM, PRRfan, PSUMark2006, PScooter63, PSzalapski, Packer333, Pan narrans, Papa528, Papushin, Parisi45, Parnell88, Patstuart, Paul August, Paul from Michigan, Paulmcdonald, Pb30, Pcpcpc, Peabody80, Pedant, PeeJay2K3, Penale52, Penfold83, Pennsylvania Penguin, Pepeeg, PerlKnitter, Persian Poet Gal, Petdance, PeterSymonds, Peterpansyndrome, Pewwer42, Phatom87, Phil Boswell, Philip Trueman, PhilipMW, PhillyPartTwo, Phyzz2, Phyzz, Pi zero, Pinkadelica, Pinkkeith, Pinteer, PlaysInPeoria, Plumlogan, Pnatt, Poindexter Propellerhead, Pokrajac, Polarbear97, Politepunk, Pollinator, Poor Yorick, Possum, PrestonH, Prolog, Proud Ho, Ptfreak, PurpleRain, Purples, Pyroclastic, Quadell, Quangbao, Quato, Quinsareth, QuiteUnusual, Qxz, R'son-W, RJHall, RJaguar3, RL1991, RLent, RMelon, Rad1986, RadicalBender, Radishes, Radon210, Radzinski, Rafimoses202, Ram-Man, Ramio457, Ranma9617, Rao11, Ratpole, RattusMaximus, Raven4x4x, Ravik, Razorflame, Rdsmith4, RealBigFlipsbrain, ReallyMale, Rebecca, Red Darwin, Red Director, Red43, RedHillian, RedWolf, RedZebra, Reddi, Redsfan10, Redsoxcool, Rehnn83, RemembertheAFL, Res2216firestar, Retired username, Rettetast, RevanFan, Revertinator, RexNL, Rhampunter99, Rhobite, Rhyfelwr, Rhysadams, Rich Farmbrough, Richard David Ramsey, Richarnd, Richiekim, Rick lay95, RickK, Rickyrab, RjLesch, Rje, Rjwilmsi, Rkstafford, Rls, Rmhermen, Rob Lindsey, Robert Merkel, Robertcowbells2, Robmc, Robsspambox, Robth, Roke, Romanm, Ron Ritzman, Ronbo76, Roodog2k, Root4(one), RossPatterson, Rotten, Rowsdower45, Royalguard11, Rozehawk, Rreagan007, Rubena, Ruedetocqueville, Rulesfan, Rupertslander, Rushingfn, RussBlau, Rwalker, Ryan Postlethwaite, Ryryrules100, Ryulong, Røed, SAUNDERS, SDMade, SFC9394, SFGiants, SGBailey, SNIyer12, SU Linguist, Sallicio, Salmar, Salt Yeung, SalvoCalcio, Salyort, Sam Hocevar, Sam Korn, Samebchase, Sannse, Saosinlampard, Sashablu, Savidan, Sbalint, Sbatchu, Sbowness, Sceptre, SchfiftyThree, SchuminWeb, Scientizzle, Scorers, Scoultas, Sdsds, Sebasbronzini, Seglea, Sekhu, Selket, Seqsea, Seraphimblade, Sergeant81, Shalllmay, Shalom Yechiel, Shane93, Shanedoick666, Shawa666, Shawnhath, Shell Kinney, Shimmin, Shinpah1, Shirulashem, Shoemaker's Holiday, Shoez, Shortmang, Shpoman, Shred-69, Siafu (usurped), Sietse Snel, Silent One, Silent Wind of Doom, SilkTork, SimonMayer, Sionus, Sjakkalle, Sjorford, SkerHawx, SkiBumMSP, Skier Dude, Skizzik, Slaad, Slicing, Sligocki, SlimVirgin, Sloneczko, Slowking Man, Slugger9062, Smalljim, Smith03, Smith120bh, Smjg, Smokedadro, Snigbrook, Snood1205, Snowolf, SoWhy, Soadaw, Solicitr, Soliloquial, Someone else, Sophitus, Soul phire, Soumyasch, Soze, Sp33dyphil, SpLoT, Spandox, Spangineer, SpeedyGonsales, Spitfire19, Splash, Splatg, Spring Rubber, Spy1986, Standleylake40, Staxringold, Steel, Stephen G. Brown, StephenBuxton, SteveCru, Stevertigo, Stevo1000, Storm Rider, Storpilot, Stumpy man, Stupid Corn, Stwalkerster, SugnuSicilianu, Sumoeagle179, Supergoalie1617, Swatjester, Swedish fusilier, SweetNeo85, Swellman, Swid, Sylent, Systemuser, T.J.V., THEN WHO WAS PHONE?, TNorthcutt, TOttenville8, Tabletop, Tangent747, Tangerines, Tangotango, Tawker, Tayger, Tbhotch, Tcncv, Teaser2k, Technopilgrim, Tedius Zanarukando, Tellyaddict, Template namespace initialisation script, Tempshill, TenPoundHammer, Tex, TexasAndroid, The Anome, The Evil Spartan, The Holy Roman Emperor, The Monster, The Moose, The Rambling Man, The Thing That Should Not Be, TheCatalyst31, TheHayMaker911, TheKMan, TheNewPhobia, TheTruthiness, TheYmode, Thebackspace, Thebdj, Thebeginning, Theda, Theiggyurracand, Themadmac, Theminddset, Thethinredline, Theword2, Thingg, Tiddly Tom, TimShell, Timc, TimeLord mbw, Timrem, Timwi, Tiptoety, Titoxd, Tiyoringo, Tj crockett, Tk508, Tleclercq, Toa Nidhiki05, Toad of Steel, Toadzilla64, Tocino, Toddst1, Toffile, Togor123, Tom harrison, Tom334, Tomtheman5, Tony James, Tony1, TonySt, TonyTheTiger, Tooga, Torc2, TornVictor, Travelbird, TravisTX, Trebor, TreveX, Trevor GH5, Trey Goette, Tricko20, Triona, Tron175, Trontonian, Trusilver, Truthanado, Trödel, Ttrain88, Tumadoireacht, Turkeyphant, Tuspm, Twas Now, Twelvethirteen, Tyler,

Lightmouse, Lights, Lir, Lopsidedman, Lordkazan, Loren.wilton, Lousyd, Lunchscale, MDR83, MLilburne, MacMac161616, Madcoverboy, Manop, Masonpatriot, Matt Keleher, Matt Yeager, Mattbrundage, Mav, Maximus Rex, Mboverload, Mdebets, Meegs, Michael Hardy, Mifter, Minnecologies, MisfitToys, Mjsabby, Moe Epsilon, Moeron, Monae08, Moncrief, Monkeybrow, Morrisbsa, Mosmof, NawlinWiki, Nbenda, Nconwaymicelli, Needlenose, Neutrality, NickVeys, Nicolaibrown, Nimitimin, NithinBekal, Noel Streatfield, Norsemark, Ohnoitsjamie, Oliver Pereira, Ommnomnomgulp, Opt 05, Ornryactor, Ottoump, P3net, Packerstud17, Parallel or Together?, Paul Stansifer, Pbroks13, PepeWheelz, Philosopher, Piano non troppo, Pinkadelica, Power, Presidentman, Punctilius, Pvsoccer20, Quidam65, R Huck, R'n'B, RTC, Racepacket, RadicalBender, RadioFan2 (usurped), Raivena, Rapiant, Rasamassen, Raymond arritt, Reedy, Reflecks, Refleks, Rgoodermote, Rhinosnore, Rich Farmbrough, Rich257, RjLesch, Rjwilmsi, Rnt20, Rodhullandemu, Rosarinagazo, Ryan Postlethwaite, RyanGerbil10, Ryanwagner, S, Saturn star, SchuminWeb, Scott Mingus, Secondarywaltz, Semperf, Seth Ilys, Shimgray, Slysplace, Smiller933, Snowolf, Spurius Furius Fusus, Ssandersisu, Steff, Steven Zhang, StevenHW, Superflush, Supermantotherescue, Superslum, Tang 12, Tansm, Tanthalas39, Tedickey, Teraldthecat, Texasguy84, The Anome, The Profressor, The REAL Pete in Pittsburgh, The Random Editor, Theman1491, Thingg, Thomasthunsen, Treydavis3, Txactor, Uncle Milty, Uofgaysian, Upholder, Vanished User 1004, Vegietales, Versus22, Vicki Rosenzweig, Victoriaedwards, Vina, Vossman, WhisperToMe, Whitejay251, WikiDon, Willking1979, Wingman1331, Wknight94, Wordbuilder, Wossi, Wrmcatee, Wtmitchell, X!, Yvesnimmo, Zirgaq, Zoe, 631 anonymous edits

**Troy_University**  *Source*: http://en.wikipedia.org/w/index.php?title=Troy_University  *Contributors*: 72Dino, Alsathletic, Anvilmedia, Atlantabravz, AuburnPilot, Autiger, BD2412, Baseball Watcher, Bessjr, BobGumble29, Bobak, Bobblewik, C4pt4in W1k1, Chris9086, D6, Dale Arnett, Datboy9in, DeFaultRyan, Disavian, DonnaSchubert, Downwards, DragoLink08, Drmies, DuncanHill, English8201, EricSerge, Evill72, Fgf2007, Finngall, Fliry Vorru, Fnlayson, Football79, Forteblast, Freddiem, Havardj, HiLo48, Highdefinition03, Hu02138, Ichabod, ImGz, Intelligentsium, JDoorjam, JNW, Janejellyroll, Jesus geek, Jgaughan, Jllm06, JodyB, John of Reading, Johnpacklambert, Jweiss11, Ken Gallager, Kreeder13, Lbr123, Levineps, LilHelpa, LindsayH, Lissoy, Lydiasenn, Mactones, Martarius, Masonpatriot, Maximus Rex, Mboverload, Meeples, Mike Selinker, Mild Bill Hiccup, Miles Montgomery, Mjj0003, Moe Epsilon, Mrtrey99, Nalvage, Naraht, Neansley, Nutmegger, Ommnomnomgulp, Patken4, RA0808, Rcreel, Reywas92, Rich Farmbrough, Richard David Ramsey, RickK, Rjd0060, RIquall, Rosem12514, Rwils, SOTSHistorian, Sacker12, Samwisep86, SebRovera, Smooth O, So God created Manchester, SportsMaster, Stealthound, Straykat99, Tassedethe, Tcncv, Thaïs Alexandrina, The Monster, The Thing That Should Not Be, TheJazzDalek, Tlatkins, Tony1, Touchmypad123, Treydavis3, Trojan1998, Troy-webmaster, Troyuni, TruckTech, Truthtellerdon'tbemad, Trythisonyourpiano, Velella, Voyagerfan5761, Zhinz, 319 anonymous edits

# Image Sources, Licenses and Contributors

file:2005PoinsettaBowl-Navy-LOS.jpg Source: http://en.wikipedia.org/w/index.php?title=File:2005PoinsettaBowl-Navy-LOS.jpg License: unknown Contributors: Journalist 2nd Class Zack Baddorf

File:Unknown Early American Football Team.jpg Source: http://en.wikipedia.org/w/index.php?title=File:Unknown_Early_American_Football_Team.jpg License: unknown Contributors: Originally published by the Detroit Publishing Company

File:Walter Camp - Project Gutenberg eText 18048.jpg Source: http://en.wikipedia.org/w/index.php?title=File:Walter_Camp_-_Project_Gutenberg_eText_18048.jpg License: unknown Contributors: ABF, Baileypalblue, Herbythyme, THEN WHO WAS PHONE?, Tagishsimon, Väsk, 11 anonymous edits

File:AmFBfield.svg Source: http://en.wikipedia.org/w/index.php?title=File:AmFBfield.svg License: unknown Contributors: Original uploader was Xyzzy n at en.wikipedia

File:NSU Football.jpg Source: http://en.wikipedia.org/w/index.php?title=File:NSU_Football.jpg License: unknown Contributors: w:User:CPackerCaleb Long

File:Nate Longshore prepares to pass at ASU at Cal 2008-10.04.jpg Source: http://en.wikipedia.org/w/index.php?title=File:Nate_Longshore_prepares_to_pass_at_ASU_at_Cal_2008-10.04.jpg License: unknown Contributors: User:CPacker

File:2006 Pro Bowl tackle.jpg Source: http://en.wikipedia.org/w/index.php?title=File:2006_Pro_Bowl_tackle.jpg License: unknown Contributors: BrokenSphere, Deadstar, Ibn Battuta, Thivierr

File:Shea Smith-edit1.jpg Source: http://en.wikipedia.org/w/index.php?title=File:Shea_Smith-edit1.jpg License: unknown Contributors: User:Jjron

File:Orton To Wolfe.jpg Source: http://en.wikipedia.org/w/index.php?title=File:Orton_To_Wolfe.jpg License: unknown Contributors: RMelon (talk). Original uploader was RMelon at en.wikipedia

File:MTSU Minnesota 2010-09-02 field goal.JPG Source: http://en.wikipedia.org/w/index.php?title=File:MTSU_Minnesota_2010-09-02_field_goal.JPG License: unknown Contributors: User:Latics

File:Second half kickoff, 2010 ACC Championship Game.JPG Source: http://en.wikipedia.org/w/index.php?title=File:Second_half_kickoff,_2010_ACC_Championship_Game.JPG License: unknown Contributors: User:Jayron32

File:Back Judge picks up flag at Rams at 49ers 11-16-08.JPG Source: http://en.wikipedia.org/w/index.php?title=File:Back_Judge_picks_up_flag_at_Rams_at_49ers_11-16-08.JPG License: unknown Contributors: User:BrokenSphere

File:American Football Positions.svg Source: http://en.wikipedia.org/w/index.php?title=File:American_Football_Positions.svg License: unknown Contributors: Kainaw on the English Wikipedia

File:Shaun Carney rushing.jpg Source: http://en.wikipedia.org/w/index.php?title=File:Shaun_Carney_rushing.jpg License: unknown Contributors: Dennis Rogers

File:AmericanFootballTraining.jpg Source: http://en.wikipedia.org/w/index.php?title=File:AmericanFootballTraining.jpg License: unknown Contributors: Nagy, TFCforever, TreveX, 1 anonymous edits

File:TurkeyBowl.jpg Source: http://en.wikipedia.org/w/index.php?title=File:TurkeyBowl.jpg License: unknown Contributors: Leemart30, Nehrams2020, 3 anonymous edits

File:Flag of Australia.svg Source: http://en.wikipedia.org/w/index.php?title=File:Flag_of_Australia.svg License: unknown Contributors: Anomie, Mifter

File:Flag of Belgium (civil).svg Source: http://en.wikipedia.org/w/index.php?title=File:Flag_of_Belgium_(civil).svg License: unknown Contributors: Bean49, David Descamps, Dbenbenn, Denelson83, Evanc0912, Fry1989, Gabriel trzy, Howcome, Ms2ger, Nightstallion, Oreo Priest, Rocket000, Rodejong, Sir Iain, ThomasPusch, Warddr, Zscout370, 4 anonymous edits

File:Flag of Brazil.svg Source: http://en.wikipedia.org/w/index.php?title=File:Flag_of_Brazil.svg License: unknown Contributors: Anomie

File:Flag of Finland.svg Source: http://en.wikipedia.org/w/index.php?title=File:Flag_of_Finland.svg License: unknown Contributors: User:SKopp

File:Flag of Germany.svg Source: http://en.wikipedia.org/w/index.php?title=File:Flag_of_Germany.svg License: unknown Contributors: Anomie

File:Flag of Hungary.svg Source: http://en.wikipedia.org/w/index.php?title=File:Flag_of_Hungary.svg License: unknown Contributors: User:SKopp

File:Flag of India.svg Source: http://en.wikipedia.org/w/index.php?title=File:Flag_of_India.svg License: unknown Contributors: Anomie, Mifter

File:Flag of Ireland.svg Source: http://en.wikipedia.org/w/index.php?title=File:Flag_of_Ireland.svg License: unknown Contributors: User:SKopp

File:Flag of Israel.svg Source: http://en.wikipedia.org/w/index.php?title=File:Flag_of_Israel.svg License: unknown Contributors: AnonMoos, Bastique, Bobika, Brown spite, Captain Zizi, Cerveaugenie, Drork, Etams, Fred J, Fry1989, Geagea, Himasaram, Homo lupus, Homo sapiens, Klemen Kocjancic, Kookaburra, Luispihormiguero, Madden, Neq00, NielsF, Nightstallion, Oren neu dag, Patstuart, PeeJay2K3, Pumbaa80, Ramiy, Reisio, Rodejong, SKopp, Sceptic, SomeDudeWithAUserName, Technion, Typhix, Valentinian, Yellow up, Zscout370, 31 anonymous edits

File:Flag of Italy.svg Source: http://en.wikipedia.org/w/index.php?title=File:Flag_of_Italy.svg License: unknown Contributors: Anomie

File:Flag of Japan.svg Source: http://en.wikipedia.org/w/index.php?title=File:Flag_of_Japan.svg License: unknown Contributors: Anomie

File:Flag of Mexico.svg Source: http://en.wikipedia.org/w/index.php?title=File:Flag_of_Mexico.svg License: unknown Contributors: User:AlexCovarrubias

File:Flag of New Zealand.svg Source: http://en.wikipedia.org/w/index.php?title=File:Flag_of_New_Zealand.svg License: unknown Contributors: Adambro, Arria Belli, Avenue, Bawolff, Bjankuloski06en, ButterStick, Denelson83, Donk, Duduziq, EugeneZelenko, Fred J, Fry1989, Hugh Jass, Ibagli, Jusjih, Klemen Kocjancic, Mamndassan, Mattes, Nightstallion, O, Peeperman, Poromiami, Reisio, Rfc1394, Shizhao, Tabasco, Transparent Blue, Väsk, Xufanc, Zscout370, 35 anonymous edits

File:Flag of Norway.svg Source: http://en.wikipedia.org/w/index.php?title=File:Flag_of_Norway.svg License: unknown Contributors: User:Dbenbenn

File:Flag of Poland.svg Source: http://en.wikipedia.org/w/index.php?title=File:Flag_of_Poland.svg License: unknown Contributors: Anomie, Mifter

File:Flag of Serbia.svg Source: http://en.wikipedia.org/w/index.php?title=File:Flag_of_Serbia.svg License: unknown Contributors: sodipodi.com

File:Flag of Spain.svg Source: http://en.wikipedia.org/w/index.php?title=File:Flag_of_Spain.svg License: unknown Contributors: Anomie

File:Flag of the United Kingdom.svg Source: http://en.wikipedia.org/w/index.php?title=File:Flag_of_the_United_Kingdom.svg License: unknown Contributors: Anomie, Mifter

File:Auburn University seal.svg Source: http://en.wikipedia.org/w/index.php?title=File:Auburn_University_seal.svg License: unknown Contributors: Avicennasis, Jalanpalmer, Neurolysis

File:Auburn University logo.svg Source: http://en.wikipedia.org/w/index.php?title=File:Auburn_University_logo.svg License: unknown Contributors: Avicennasis, Jalanpalmer, Neurolysis

Image:1883 Old Main Building South College Street Auburn Alabama.jpg Source: http://en.wikipedia.org/w/index.php?title=File:1883_Old_Main_Building_South_College_Street_Auburn_Alabama.jpg License: unknown Contributors: Auburn University digital scan of photograph taken by unknown photographer

Image:1890s Samford Hall Auburn Alabama.jpg Source: http://en.wikipedia.org/w/index.php?title=File:1890s_Samford_Hall_Auburn_Alabama.jpg License: unknown Contributors: Auburn University digital scan of photograph taken by unknown photographer

Image:Auburn cadets.jpg Source: http://en.wikipedia.org/w/index.php?title=File:Auburn_cadets.jpg License: unknown Contributors: Original uploader was Kiranlightpaw at en.wikipedia

File:Auburn University Athletics logo.svg Source: http://en.wikipedia.org/w/index.php?title=File:Auburn_University_Athletics_logo.svg License: unknown Contributors: FleetCommand

Image:AuburnUniversity-SamfordHall.jpg Source: http://en.wikipedia.org/w/index.php?title=File:AuburnUniversity-SamfordHall.jpg License: unknown Contributors: Original uploader was Autiger at en.wikipedia

Image:Ross Chemical Laboratory - Auburn University - IMG 2799.JPG Source: http://en.wikipedia.org/w/index.php?title=File:Ross_Chemical_Laboratory_-_Auburn_University_-_IMG_2799.JPG License: unknown Contributors: User:Daderot

Image:AuburnAthleticBuildings.jpg Source: http://en.wikipedia.org/w/index.php?title=File:AuburnAthleticBuildings.jpg License: unknown Contributors: User:AuburnPilot

Image:Aubie-01.jpg Source: http://en.wikipedia.org/w/index.php?title=File:Aubie-01.jpg License: unknown Contributors: AuburnPilot, JGlover, Merbenz, Spyder Monkey, Wareagle2012, Ytoyoda, 1 anonymous edits

Image:Auburn Tigers.jpg Source: http://en.wikipedia.org/w/index.php?title=File:Auburn_Tigers.jpg License: unknown Contributors: Original uploader was Organizedchaos02 at en.wikipedia

Image:Jordan hare stadium.jpg Source: http://en.wikipedia.org/w/index.php?title=File:Jordan_hare_stadium.jpg License: unknown Contributors: Original uploader was AuburnPilot at en.wikipedia

Image:AuburnWatertower.png Source: http://en.wikipedia.org/w/index.php?title=File:AuburnWatertower.png License: unknown Contributors: Original uploader was Porsche997SBS at en.wikipedia (Original text : Porsche997SBS (talk))

Image:AmericanFootball current event.svg Source: http://en.wikipedia.org/w/index.php?title=File:AmericanFootball_current_event.svg License: unknown Contributors: User:Anomie, User:Tkgd2007, User:Zzyzx11

Printed by Books on Demand GmbH, Norderstedt / Germany